Meet the editor

Dr. Hoda S. El-Sayed is an Associate Professor of Dairy and Food Microbiology in the Dairy Department, at the National Research Centre, Egypt. She graduated from the Faculty of Agriculture, Cairo University, where she received her MSc and Ph.D. in 2008 and 2013, respectively. She has published fifty-five research articles on the microencapsulation of probiotics, the development of functional products, controlling the shelf life of food using plant extracts, and the use of nanotechnology in food packaging. She has one Egyptian patent to her credit. Dr. El-Sayed received the award for best scientific paper from Unilever Mashreq in 2016, the Scientific Encouragement in Dairy Science award from the Egyptian Society of Dairy Science in 2017, and a best scientific paper award from the Dream Company in 2019 and 2020.

Contents

Preface

In recent years, the microbiome and its impact on health has become an area of great scientific interest. There are many microorganisms present in the human body, and these are mainly located in the gut. Generally, there are two types of microorganisms in the gastrointestinal tract: "good" and "bad." Good microorganisms are called probiotics, which are living microbiota cultures that enhance the qualities of the indigenous microbiota when supplied to the host. These bacteria are widespread in nature and are suitable for usage in food industries. *Lactobacillus* is one of the earliest probiotics proven to have a beneficial impact on health. Probiotics enhance epithelial barriers, increase adherence to gut mucosa and microbial adhesion, generate antimicrobial compounds and regulate the immune system, and can be used as food supplements to manage several gastrointestinal tract diseases. They are also used in research studies to develop commercial probiotic foods. Therefore, a perfect and sensitive balanced interaction of microbes with the host is required for a healthy body. Any disturbance in that balance leads to dysbiosis and the host may become more susceptible to disease.

This book discusses the effect of microbiota on human health. Microbiota is the powerhouse of health and disease. Moreover, the microbiome contains the genetic information and the genomes of the microorganisms themselves. It is now well known that the microbiome interacts with its host and is involved in basic human biological processes, modulating the metabolic phenotype in the bioconversion of nutrients and detoxification, influencing innate immunity, and protecting against microbial infections. The microbiome is also known for producing many vitamins such as vitamin B12, thiamine, riboflavin, and vitamin K, which is required for blood coagulation. Generally, the gut microbiota construction is determined by several factors, including gestational pathologies, type of birth, type of feeding, prenatal and perinatal use of antibiotics, complementary feeding, and environmental pollutants. From gestation to the first two years of life, these events influence the establishment of the microbiota. Hence, microbiota affects the metabolic and immune response and has a subsequent impact on human health.

As an alteration in the microbiome can be protective or causative, this book reviews the pathogenesis and potential roles of some members of microbiota in diseases such as inflammatory bowel disease (IBD) disease and presents a promising strategy to alleviate and cure this condition. It also examines the relationship between gut microbiota imbalance and how defects in this dysbiosis can lead to disease. Moreover, the book discusses microbiota potential in type 2 diabetes, highlighting recent findings in the regulation of extraintestinal metabolism by gut microbiome with emphasis on the physiology and pathophysiology of the pancreas in health and disease. The microbiome also plays a major role in the development of obesity by regulating energy metabolism. The makeup and density of intestinal flora can be influenced by diet. As such, this book examines the relationship between the gut microbiome and obesity. It also examines the role of gut microbiota in promoting the development and progression of brain health. The interactions of the gut microbiota and brain axis have been studied using various animal models. However, most of the animal research has only been able to reveal the fundamentals, such as the diversity of the microbial community, the potential microbial pathways, and the dysbiosis of the gut microbiota due to diet and drugs. Finally, the book reviews the development of functional food fortified with probiotic microorganisms, such as fermented dairy products (yogurt, fermented beverages, and others), which have a positive impact on the gut microbiota balance. Probiotic fermented milk should contain at least 10^7 CFU/mL of live bacteria at the time of consumption to obtain health benefits.

Hoda El-Sayed
Associate Professor,
Dairy Department,
Food Industries and Nutrition Research Institute,
National Research Centre,
Cairo, Egypt

Section 1

Microbiota and Health

Chapter 1

Microbiome - The Power House of Health and Disease

Basavaraju M., Gunashree B.S. and Srinath B.S.

Abstract

The field of microbiome is an exciting and rapidly expanding research over the past few decades that have become a topic of great scientific and public interest. Microbiome comprises a complex collection of microorganisms, with their genes and metabolites colonizing different body niches in a deep symbiotic relationship in the aspect of both health and diseases. Microbial populations vary across the body sites, driven by different environmental condition, immunological factors and interactions between microbial species. It is now well known that the microbiome interact with their host, assisting in the bioconversion of nutrients and detoxification, boosting immunity and protecting against pathogenic microbes, maintaining individuals' health. A wide range of environmental factors can have an impact on gut microbiota imbalance, which has a strong link to health and disease. The microbial role in basic biological processes as well as the development and progression of major human diseases like infectious diseases, liver diseases, gastrointestinal cancers, metabolic diseases, respiratory diseases, mental or psychiatric diseases, and autoimmune diseases. Therefore, a perfect and sensitive balanced interaction of microbes with the host is required for a healthy body. With recent advances in genome sequencing and 'meta-omics' tools, culture-independent analyses of microbiomes have been made possible, thus accelerating the progress of microbiome research by leaps and bounds.

Keywords: data analysis, dysbiosis, microbiome, metagenomics, microbiota, probiotic

1. Introduction

Microbes inhabit almost all human body parts and play a critical role in human health and disease. Research has increasingly focused on the diverse microbial communities that interact with the host to influence disease processes as modern microbiology and next-generation sequencing technologies have evolved. The term 'microbiome' refers to the complex blend of microorganisms such as bacteria, bacteriophage, viruses, fungi, single-celled animals and their genes as well as metabolites. Colonizing different body niches which contribute in big ways to human health and wellness. As microbial communities, also known as the microbiota, microorganisms, or microbes, coexist and interact with one another and with the surrounding environment. The microbial communities within our body are highly personalized and considered as unique to each individual as their fingerprints [1] also unique to each body sites [2]. This can also be referred to as the metagenome of the microbiota. The word "microbiome" was coined by Joshua Lederberg, who was the first to use it to "symbolize the ecological community of commensal, symbiotic, and pathogenic microorganisms that literally occupy human body space and have been largely overlooked as health and disease determinants" [3]. Over time, the term microbiome has evolved not only to refer microbiota, but also to the genetic information and the genomes of the microorganisms themselves. It is now well known that the microbiome interacts with its host and also involved in basic human biological processes, modulating the metabolic phenotype in the bioconversion of nutrients and detoxification, influencing innate immunity and protecting against microbial infections. Microbiota boosts the immune system, breaks down potentially hazardous dietary molecules, and synthesizes vitamins like vitamin B12, thiamine, riboflavin, and vitamin K, which is required for blood coagulation [4]. Microbiome is home to trillions of symbiotic microorganisms in which some of these are useful, and some are harmful and it supports many physiological functions, helps in maintaining the integrity of our gut lining, and protects us from disease and infection. Therefore, a perfect and sensitive balanced interaction of microbes with the host is required for a healthy body. The microbiota has many more metabolic genes than the human genome and provides unique enzymes and biochemical pathways to humans [5]. Furthermore, many of the positive metabolic macrobiotic activities for the host are engaged in either food acquisition or xenobiotic processing, such as the metabolism of undigested carbohydrates and vitamin production [6].

Second, through competitive exclusion and the generation of antimicrobial compounds, the human microbiota acts as a physical barrier, protecting its host from invading pathogens [7].

Understanding how microbial metabolites influence the health or disease status would have a significant impact on treating diet related diseases [8]. Microbes that cause disease build up over time, affecting gene activity and metabolic processes and causing an incorrect immune response to substances and tissues that are normally present in the body. Autoimmune diseases tend to be passed down through generations via microbiome inheritance rather than DNA transmission [9]. Recent studies revealed that the associated microbes stimulates the normal development of the humoral and cellular mucosal immune systems and the signals and metabolites of microorganisms can be sensed by the hematopoietic and non-hematopoietic cells of the innate immune system and translated into physiological responses [10]. Often, reduction in microbial diversity and outgrowth of specific species can induce negative effects like inflammation or infection [11]. However, most of the microbial taxa and species of the human microbiome are still unknown. Without revealing the identity of these microbes as a first step, we cannot appreciate their role in human health and diseases [12].

There are plenty of projects trying to decode the human genome by sequencing all human genes. In a similar way, the microbiome has been subject to intensive efforts to unravel all its genetic information. Advances in omics-based techniques have contributed to a better knowledge of the microbiome and the many factors that influence its microbial composition. Understanding the entire spectrum of the "microbiome's" role in health and disease is still in its infancy. Our bacterial flora clearly plays a far larger influence in systemic disorders than previously thought [13]. High throughput sequencing reveals the amazing complexity and extent of the microbial communities that reside within or upon us therefore various computational approaches are available to analyze the microbiota on an unprecedented scale [14]. Recent scientific advances in genetics mean that humans know a lot more about the microbes in the body. Researchers from across the globe are investigating how changes in the microbiome are linked to, or perhaps cause, illnesses, as well as developing new therapeutic ways to modify the microbiome to cure disease and restore and support health. In addition, microbiome research is gaining tremendous interest as documented by the explosion in publications with more than 20,000 articles published in 2020 alone. The rapid development of new molecular tools such as transcriptomics, metagenomics, and metabolomics has aided in the recent advancement of microbiome results linked to humans. These fast evolving recent technologies are enhancing our ability to comprehend the human body and the microbiome that affects health. Researchers need to conclude with future directions and how to convert the basic science into translational medicine and development of innovative microbiome-based therapy.

2. Microbiome's evolution

Microbiome are the home tract of wide range of microorganisms that can be commensal, symbiotic, or toxic to all multicellular organisms, including plants. The microbiota includes bacteria, archaea, protists, fungi, and viruses, all of which have been shown to be vital for their host's immunologic, hormonal, and metabolic balance [15]. Microbial communities live in multiple body sites in humans and animals (including the stomach, oral cavity, esophagus, skin, and vagina) and interact with and influence their hosts' immune system and metabolism. In addition, microbes have developed alongside humans and are now an essential component of life, performing a variety of essential roles. Due to changes in environmental parameters such as temperature, pH, oxygen, and nutrition availability, their composition varies greatly between body locales and specific biogeography. Although much has been done to explore its diversity, a full understanding of our microbiomes demands an evolutionary perspective. At the strain level, microbial evolution may occur (e.g., when advantageous mutations in specific genes drive adaptation to new selection pressures) selection may also enhance the frequency of a specific microbial taxon, causing the adaptive microbiome's microbial taxa to be lost [16]. The microbiome can evolve at two levels: first, each individual microbe is subjected to evolutionary processes (mutation, selection, migration, drift, speciation, etc.), and second, a host species' microbiome can evolve by incorporation and elimination of microbial taxa, or by changes in their relative abundances as a consequence of these evolutionary processes [17].

Interestingly, mammals that have independently evolved on herbivorous diet often exhibit similar microbiomes [18]; however, this is not the case of panda bears, whose microbiome resembles that of their carnivorous and omnivorous close relatives, despite the panda's herbivorous diet, probably due to phylogenetic constraints [19]. The compositional overlap between the gut microbiota of species populations in the western hemisphere correlates with their geographic proximity in most mammals, and each geographic location has a distinct microbiome composition that is not attributable to the diets or evolutionary histories of the mammals living there, suggesting that horizontal transmission also shapes the microbiome [20]. Because one species and its associated microbiome serve as the meal for the paired predator, this link is most visible in sympatric predator-prey groups. The structure of the relationships in primate species is unknown, but they are likely to follow some of the same patterns.

It is important to remember that the microbiome is a complex and dynamic ecosystem and multiple overlapping factors shape the microbiome composition and it is unique in each individual, and the differences among individuals are largely compared to the typical biochemical differences within a person over time. The gut microbiota is shaped by a variety of factors, including genotype, dietary composition and mode of delivery, recreational drugs, antibiotic therapy, pre and probiotic treatment, lifestyle (e.g., smoking and physical activity), social interactions, and environmental exposure to various xenobiotics. In addition, several other factors are also involved including (i) Diet. The types of food that a person consumes can have a significant impact on gut microbiota. (ii) Exposure to pathogens, (iii) Age, (iv) Psychological Stress/Anxiety, (vi) Medication/Drug Use, (vii) Tobacco Use, and Alcohol Consumption (vii) Physical Activity [21]. One important factor emerging from the research advances is the importance of microbial diversity. In healthy settings, an individual's microbiota is more diverse than in sickness, when diversity is diminished. Low microbiome diversity has been linked to metabolic inefficiency, skin issues, gastrointestinal problems, and low-level inflammation.

Because of the biological interaction of the organisms with the immune system throughout time, the indigenous organisms in the human body are well adapted to the immune system. A shift in the gut microbial flora plays a crucial impact in human health and disease pathogenesis. These changes are caused by a combination of factors, including lifestyle and the existence of an underlying disease. Dysbiosis makes the host more susceptible to infection, the type of which varies depending on the anatomical place. The precise metabolic activities and functions of these microorganisms within each bodily location are accounted for by the inherent diversity of the human microbiota. As a result, it's critical to comprehend the human microbiome's microbial composition and behaviors as they relate to health and disease. The microbiome can affect many physiological processes in our body, including immune system development, the ability to process dietary polysaccharides, vitamin and hormone production, pH regulation, processing and detoxification of environmental chemicals and maintenance of the skin and mucosal barrier function [22, 23]. There has been a boom of research into how the microbiota of the gastrointestinal system affects human health and disease, and what treatments might be made, particularly in the last decade (**Table 1**).

Phylum	Class	Characteristics	Examples
Firmicutes	Bacilli; Clostridia	Gram-positive bacteria with a variety of morphologies (rod, coccoid, spiral) and physiologies (anaerobic, aerobic); commensal and helpful bacteria.	*Lactobacillus*; *Ruminococcus*; *Clostridium*; *Staphylococcus*; *Enterococcus*; *Faecalibacterium*
Bacteroidetes	Bacteroidetes	Gram-negative; made up of three main classes that are widely spread in the environment, such as soil, ocean, and animal intestines.	*Bacteroides*; *Prevotella*
Proteobacteria	Gammaproteobacteria; Betaproteobacteria	Gram-negative; include a wide variety of pathogens	*Escherichia*; *Pseudomonas*
Actinobacteria	Actinobacteria	Gram-positive; diverse morphology; major antibiotic producers in the pharmaceutical industry	*Bifidobacterium*; *Streptomyces*; *Nocardia*

Table 1.
In the human body, the most common bacterial phylum [24].

3. Human microbiome project

The major goal of the human microbiome project is to define the number, diversity and functionality of genes found in all bacteria that live in various parts of the human body on a permanent basis and analyze its role in human health and disease. The gut microbiota expresses around 3.3 million bacterial genes, compared to only 20,000 genes in the human genome. Studies show that manipulating non-pathogenic bacterial strains in the host can help the immune system recover from disorders caused by pathogenic bacteria. An ever-growing number of studies have demonstrated that changes in the composition of our microbiomes correlate with numerous disease states, raising the possibility that manipulation of these communities could be used to treat disease. The microbiome of a person can affect their susceptibility to infectious diseases and contribute to gastrointestinal chronic disorders including Crohn's disease and irritable bowel syndrome. A person's response to a pharmacological therapy is determined by a group of microorganisms. The mother's microbiome may have an impact on her children's health.

Researchers researching the human microbiome are discovering previously unknown organisms and genes all around the world. Various combinations of microbial species have been related to certain human health issues in genetic studies that quantify the relative abundance of different species in the human microbiome. A thorough understanding of the diversity of microbes in the human microbiome could lead to new therapeutics, such as producing more "good" bacteria to cure a bacterial infection caused by "bad" bacteria. The HMP is a road plan for understanding and describing the role of the microbiome in health, nutrition, immunology, and disease.

4. Microbiota benefits of the body

The microbiome is essential for human development, immunity and nutrition. Microbiota boost the immune system, break down potentially harmful dietary components, and manufacture vitamins and amino acids such vitamin B and vitamin K [25]. The major enzymes required for the formation of vitamin B12 are exclusively present in bacteria, not plants or mammals [26]. Bacteria living in and on the human body are not always invaders but beneficial colonizers too. Sugars like table sugar and lactose (milk sugar) are quickly absorbed in the upper portion of the small intestine, while more complex carbs like starches and fibers are more difficult to digest and may end up in the large intestine. By creating digestive enzymes, the microbiota aids in the breakdown of these substances. Short chain fatty acids (SCFA) are produced when indigestible fibers are fermented, and they can be utilized by the body as a food source as well as play a role in muscular performance and possibly the prevention of chronic diseases including cancer and bowel disorders. SCFA has been demonstrated to be effective in the treatment of ulcerative colitis, Crohn's disease, and antibiotic-associated diarrhea in clinical trials [25].

Autoimmune diseases like diabetes, rheumatoid arthritis, muscular dystrophy, multiple sclerosis, and fibromyalgia have been linked to microbiota dysfunction. Microbes that cause disease build up over time, altering gene activity and metabolic processes, leading in an aberrant immune response to chemicals and tissues that are usually present in the body. A healthy person's microbiota will also defend them from harmful organisms that enter the body by drinking or eating polluted water or food such as

Prevotella, *Ruminococcus*, *Bacteroides*, and *Firmicutes* are large families of bacteria found in the human stomach. Anaerobic bacteria such as *Peptostreptococcus*, *Bifidobacterium*, *Lactobacillus*, and *Clostridium* can be found in the colon due to the low oxygen environment [27]. These microbes are thought to prevent harmful bacteria from overgrowing by competing for nutrition and attachment sites on the mucus membranes of the gut, which are a significant site of immune activation and antimicrobial protein production [28, 29]. Autoimmune diseases appear to be passed in families not by DNA inheritance but by inheriting the family's microbiome. Recent studies on gut microbiota modulation suggest that probiotics should be used in the treatment of patients with severe COVID-19 infection, according to the National Administration of Traditional Chinese Medicine and China's National Health Commission [30]. Probiotics are used to prevent secondary bacterial infection and maintain intestinal microbiota balance.

5. Microbiome analysis techniques

Microbiome research is a highly transdisciplinary field with a wide range of applications and methods for studying it. There are a number of different technologies available to study the microbiome. Traditional microbiology has historically focused on the study of individual species as isolated units. In the mid-2000s, advances in DNA sequencing technology spawned a new branch of study known as

Figure 1.
Microbiome-researching technologies [32].

metagenomics, which allows for a comprehensive exploration of microbial communities without the requirement for culture. Instead of looking at the genome of a single bacterial strain cultivated in a lab, the metagenomics approach looks at a collection of genomes derived from microbial communities collected in natural settings, providing new insight into the complexity of human microbial populations [31] (**Figure 1**).

The identification of about 70% of human microbiota, which was not possible by the existing conventional microbiological methods, has been made possible by the development of the advanced techniques of metagenomics, metatranscriptomics, and metabolomics [33]. Metagenomic is a biotechnological perspective of studying the genome structure of the DNA directly extracted from their natural source [34]. Scientists have utilized these revolutionary approaches to prove the existence of genes from over a thousand different microbial species in our bodies. The metagenomic technique has the potential to uncover novel genes, gene families, and their encoded proteins that could have major implications in biotechnological and medicinal research. It enables us to investigate the makeup of a microbial population [35] (**Figure 2**).

Currently, multiple multinational organizations such as the HMP project and various other independently functioning programs are constantly generating huge amounts of data relating to metagenomic studies, and their microbiome data collection is managed by the Genomes Online Database (**Table 2**).

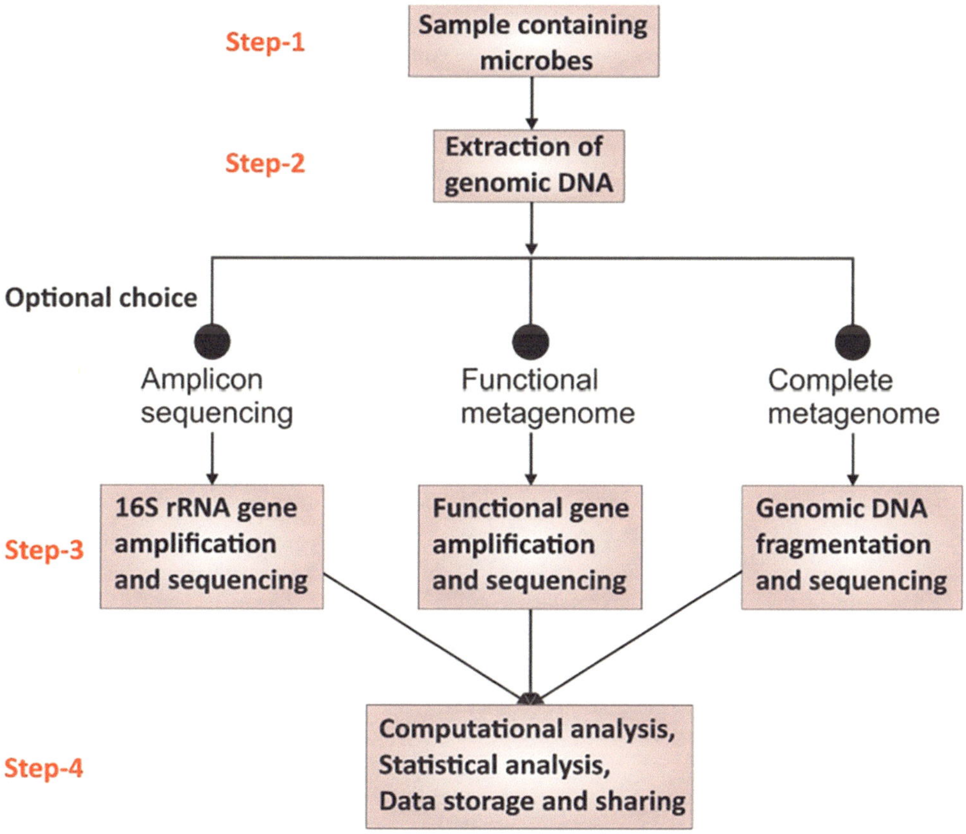

Figure 2.
Major steps in the most widely acknowledged genomics strategy for human related microbiome studies are depicted graphically [2].

Methods	Description	Advantages	Limitations	References
Models that are free of germs (GF)	Transplant germ-free in vitro embryos into germ-free moms and rear animals without coming into contact with microorganisms.	To test previously reported relationships, a "blank state" approach was deployed.	Differences in regional makeup are challenging to capture in a compromised, expensive technology that is not reflective of natural microbiome functioning.	[37, 38]
Human sampling	Population is divided into sub groups based on specific characters	Cost effective and relatively easy to access body sample sites	Distinction in regional composition are difficult to capture	[39, 40]
Population scale	Involves sampling from a selected large group of individuals	Large scale conclusion can be drawn, with broadly applicable results	Diversity within individual microbiomes is not considered with purely association based results	[41]
In vitro modeling	Experimental laboratory systems mimicking processes occurring within a living organism	Enables examination of relationship between specific microbes and host	System lacks host level complexity due to reduced microbial communities and simplified environmental structuring	[42]
Patterning of co-occurrence networks	Investigate the impact of organisms and environmental factors on community interactions.	Microbe-microbe interactions and their relationships can be examined to establish ecological network components within microbiomes.	The complexity of the microbial community is reduced, resulting in simpler system operation.	[43]
Direct observation via fluorescence	Probe specific sites or organismal components such as cells, allowing microscopic observation	Taxonomy locality and community organization can be evaluated and screens for specific phenotypes are possible	Photo bleaching can occur	[44]
Bioinformatics	Use of software tools to understand biological data, especially with large complicated data sets	Allows for rapid organization and analysis of data	Often expensive, while drawing association based conclusions	[43, 45]
Association studies	Identify genes correlated with disorders	Can discover correlative relationships between microbes and their hosts	The mechanisms and causative factors underlying correlations remain unknown	[46, 47]

Methods	Description	Advantages	Limitations	References
Meta-omics	Include metagenomics, metatranscriptomics, metaproteomic and metabolomics data collection	Analyze and detect molecular and genetic components and mediators and metabolic profiles	Equipment is highly sensitive and expensive, limiting reproducibility.	[48]
Machine learning models that predict the future	Algorithms are used to find patterns and behavior in datasets.	Use the ease of in situ analysis to find connections between microorganisms and variables.	With association-based and time-consuming data collecting, it's difficult to capture the intricacy of individual microbiomes.	[49]

Table 2.
Methods for analyzing the microbiome [36].

6. A mechanistic link between human health and disease and the microbiome

The microbiome can take up to 40% of our weight and can do many things. The human body is home to a microbiome, which is a networked community of microbes that outweigh the body's own cells. The human microbiome has piqued researchers' interest in recent years due to the microbiome's deep ties to human health. The human microbiome, also known as "our second genome," has developed alongside humans for millions of years and plays an important role in human health. Understanding the human microbiome's composition and function can help us better comprehend its structural and functional features. Understanding the human microbiome and applying metagenomic analysis to specific individuals will considerably improve our understanding of human health and diseases in the future. The study of the human microbiome and metagenome is seen as a new frontier in human genetics.

The majority of study on the human microbiome has focused on the microbes that colonize the human digestive system, as these microbes are thought to have a variety of effects on human health. The digestive system's microbiome is extraordinarily varied, with significant differences in its contents between individuals [50]. Extraneous variables, such as fecal transplantation and dietary intervention, have been proven to modulate the microbiome, which has been shown to be a viable therapeutic method to addressing a variety of health-related disorders [51]. The gastrointestinal tract (GIT) is home to a diverse range of microorganisms, which are connected by microbe-microbe and host-microbe interactions [52]. Microbial guilds (species that share resources) have been discovered to have intriguing traits that can help researchers better understand processes at both the single cell and community levels. Microbes are commensal and mediate digestion, enhance the immune system, and inhibit or prevent infections from penetrating the body under normal physiological conditions. The relationship between the human microbiome and human health is still largely unknown and unexplored, but a decrease in the diversity of the digestive system microbiota has been linked to diseases such as eczema [53], asthma, and inflammatory diseases [54], diabetes and obesity [55], allergies [56], and digestive tract disorders such as IBD (inflammatory bowel disease) [57], and IBS, according to a number of epidemiological studies (irritable

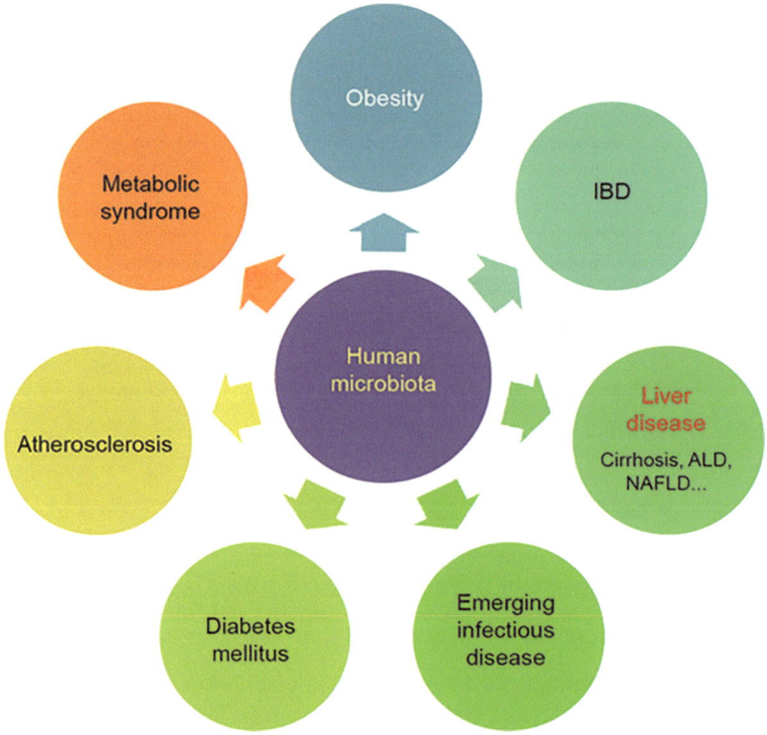

Figure 3.
Human microbial symbiosis has a close relationship with diseases of different systems [68].

bowel syndrome) [58]. Chronic fatigue syndrome [59], cancer [60], colitis [61], and bacterial vaginosis [53, 54] have all been linked to dysbiosis (microbial imbalance). A number of recent studies have shown the importance of the gut microbiome in modifying immunological responses, including immune tolerance, via Treg (T regulatory) cell modulation. Short-chain fatty acids (SCFA) have been shown to increase the formation of Treg cells in the gut, according to Geuking et al. [62]. Microbes that live in the gut aid in the breakdown of complex carbohydrates and the usage of polysaccharides [5, 63]. Other health-promoting roles of the gut microbiome include immunological regulation [64], fecal microbiome transplantation [65], metabolism, xenobiotic toxicity, and pharmacokinetics, to name a few [66]. Therefore, patients with respiratory infections and diseases were shown to have gut dysbiosis and concomitant problems, showing gut-lung crosstalk, this phenomenon can also be seen in COVID-19 patients. As a result, boosting gut microbiota using probiotics and other beneficial bacteria is significant in therapeutic applications, and this could be extended to COVID-19 treatment as a new therapeutic approach according to Srinath et al [67] (**Figure 3**).

7. The function of microbiome in terms of human health

When you realize that there are as many microorganisms in the body as there are human cells, the microbiome's importance seems understandable. The human microbiome is diverse at each body site, such as the gut, skin, mouth, and nasal cavities, where each community of microorganisms is unique. The core microbiome of a person is developed during the first years of life, although it can alter over time as a result of several factors such as nutrition, drugs, and environmental exposures. Individual vulnerability to various diseases may be determined by differences in the microbiome, which may lead to varying health outcomes from environmental exposures. A healthy microbiome has been found to play a significant role in maintaining good health [69]. Environmental exposures can also alter a person's microbiome, thereby increasing the risk of acquiring diabetes, obesity, cardiovascular and neurological illnesses, allergies, and inflammatory bowel disease. The human microbiome is primarily concentrated in the stomach. These organisms serve a critical role in maintaining and preserving human health. Previous research on the human microbiome project has shown that alterations in the immunological environment can be connected to a dysbiotic gut flora. Dysbiosis has also been related to life-threatening health disorders such as cancer, cardiovascular disease, bowel inflammatory disease, and difficult-to-treat bacterial infections due to antibiotic resistance [70]. A healthy microbiome is a diverse and abundant one, and everything from our nutrition to our surroundings influences how effectively it performs.

However, antibiotic usage and ultra-processed food consumption, for example, are destroying our gut microbiota, making people more susceptible to infections such as *Clostridium difficile* and other diseases. It's only now becoming obvious how important the link between our microbiomes and our health is. The revelation that we can use our microbiome to help us treat or even prevent disease has been perhaps the most significant development. In the last 20 years, the advent of hyper virulent *Clostridium difficile* strains has resulted in a massive increase in infections, with over 20% of cases

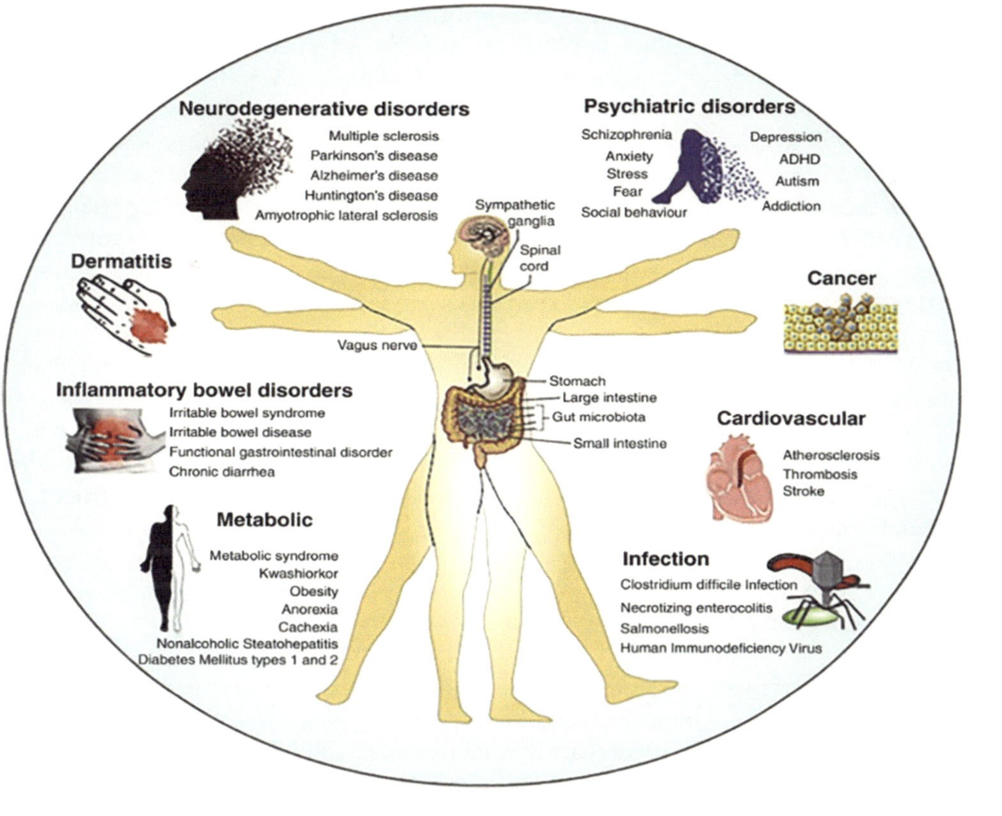

Figure 4.
Schematic diagram showing the microbiome implicated in a variety of disorder, including skin, metabolic, and cardiovascular disease, as well as cancer, infection, and neurodegenerative and psychiatric disorders [72].

now involving germs that are drug resistant. The human genome consists of just 23,000 genes, while the microbiome as a whole contains many millions. Scientists are also looking into using microbiota transplants for a wide range of disorders, such as insomnia, Parkinson's disease, HIV, chronic fatigue, multiple sclerosis, obesity, insulin resistance, metabolic syndrome, and autism [71] (**Figure 4**).

8. Antimicrobial resistance

Microorganisms must discover an optimal strategy to coexist with other microbes in a restricted environment in order to thrive. Microbes compete with one another in their surroundings for limited nutrients and space. As a result, they've devised strategies to regulate their own needs by interacting with other microorganisms. Producing antimicrobial compounds that can hinder or kill another germ is one effective technique to do so. Some microorganisms, on the other hand, have evolved mechanisms to survive in the presence of naturally occurring antimicrobials, allowing them to remain a stable member of a microbial community. Antimicrobial resistance can be inherent or acquired in microorganisms. Intrinsic resistance occurs when a bacteria develops resistance to an antibiotic on its own. Microorganisms have had intrinsic resistance mechanisms for millennia, promoting their co-evolution and integration with microbial communities. Antimicrobial medication development and use to treat and eradicate microbial infections is without a doubt the greatest triumph in contemporary medicine. Penicillin, the first mass-produced antibiotic used on a massive scale around the world, saved millions of lives and paved the way for the discovery and development of hundreds of different antimicrobial medications to combat specific infections. Antimicrobial medications have all come from naturally occurring microbial sources, to which certain microorganisms had already evolved innate resistance.

Microbes have developed acquired resistance to antimicrobial medications as a result of increased use of antimicrobial drugs combined with pre-existing resistance. Antimicrobial resistance (AMR) arises when bacteria, viruses, and fungi grow resistant to antibiotics. As a result, infections may become more difficult to remove. AMR is now considered to be one of the most serious risks to world health, food security, and economic development. According to the World Health Organization, at least 700,000 people die each year from drug-resistant diseases, and this number is expected to climb if adequate interventions are not implemented. The overuse and misuse of antimicrobial therapies in a fast rising global economy and population has resulted in a rise in the rate of AMR cases over the last 20 years. Antimicrobials, which are thought to be a panacea for eradicating illnesses, have fueled the emergence of antimicrobial resistance in bacteria.

9. Finding biomarkers in microbiome research

These types of mechanistic tests are currently being carried out in humans by several investigations. The authors assessed the ability of the individual's blood to create cytokines following several antigen challenges in 500 European-ancestral individuals in the Netherlands, and then linked this with data from their gut metagenome. According to the findings, the yeast *Candida albicans* had a particularly strong influence on the host's TNF-alpha response [73]. These investigations are particularly relevant when dealing with persons who have naturally occurring genetic knockouts or variant alleles. As has been proven for Parkinson's disease, these human genetic variants may enable microbially caused disease that may be investigated in mice with analogous null or variant genetic changes [74].

Characterizing microbial biomarkers offers a lot of promise for precision medicine, and it's a straightforward method to get microbiome research into clinical practice. For example, we know that bacterial probiotics (living bacteria purposely introduced to an animal to have a therapeutic effect) can be utilized to augment immune checkpoint blockade therapy for melanoma patients based on landmark animal studies [75]. Microorganisms in the gut have been identified as biomarkers for diagnosis that can predict if patients are at risk of developing checkpoint blockade therapy after studying the microbiomes of melanoma patients prior to immune checkpoint blockade medication.-colitis caused by a blockage [76].

These prospective studies are critical for correlating the structure, function, and metabolic products of microbial communities to health consequences. Many ongoing investigations, such as the National Institutes of Health Common Core program. Environmental Influences on Child Health Outcomes (ECHO: https://www.nih.gov/echo), now provide the infrastructure to sequence healthy, susceptible, and diseased participants to examine how lifestyle and environmental experiences shape the development of immune, endocrine, and neurological conditions. Although single time point investigations of birth cohorts show fascinating statistical relationships [77], longitudinal prospective studies accompanied by mechanistic tests in animal models are needed to determine if a specific microbiome causes disease (**Figure 5**).

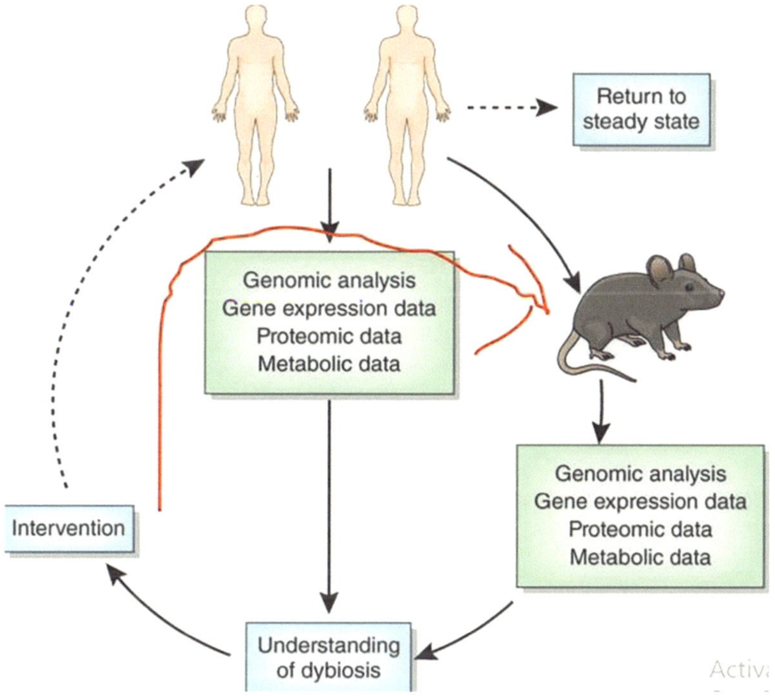

Figure 5.
The iterative cycle of analysis, interpretation and translational intervention that facilitate moving microbiome research out of correlative observation and into therapeutic treatments [78].

10. Microbiome applications

Studying the human microbiome is helping researchers to understand how the body responds to different diets, diseases and drugs. The human microbiome can be seen of as a source of genetic variety, a disease modifier, a significant component of immunology, and a functional entity that regulates metabolism and modulates drug interactions. On the one hand, there are numerous possible probiotics or helpful bacteria that could help to prevent or treat various diseases, albeit most of them are now unavailable for cultivation [79]. Little was known about the variety of microorganisms that happily dwell inside and on our bodies more than a decade ago, but researchers today believe they have the potential to influence the future of human health and examining linkages between health and disease. Almost 70% of the bacteria that make up the human microbiota are uncultivable, and many of them are anaerobic (so can only be cultivated without oxygen). These barriers have prompted researchers to investigate meta-genomics and in vivo models. While in vitro models of the digestive tract can be used to simulate one or more stages of digestion (in the stomach, small intestine, or colon), they are still incapable of duplicating the complexity of host-microbiota interactions [80]. Therefore the recent scientific evidence suggests that a healthy and diverse microbiome is beneficial to human health and the microbiome is becoming a cornerstone of preventive medicine (**Figure 6**).

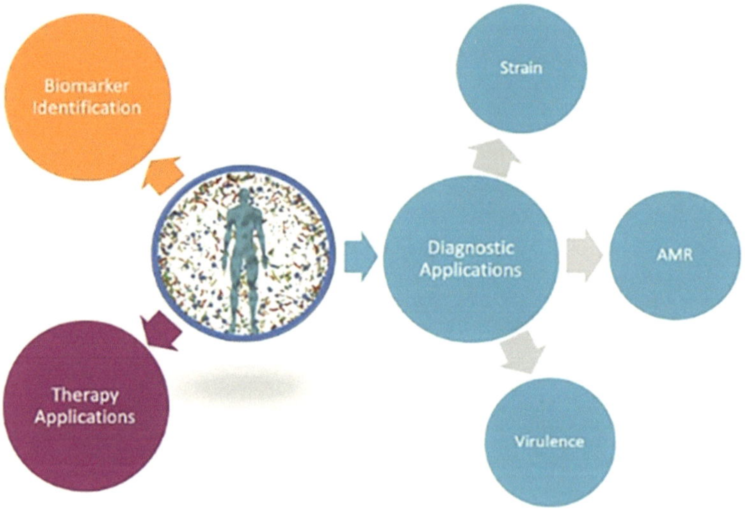

Figure 6.
The microbiome can be used to study, diagnose and treat diseases.

11. Conclusion

Microbiome research has made great progress in the past decades due to recent scientific advances in genetics and genomics. Consequently, despite being a relatively new field, microbiome has been successfully employed to alter microbiota and demonstrate promising prospects for therapeutics. It's worth noting that long-term usage of broad-spectrum antibiotics has the potential to damage the human microbiome. As a result, the indigenous microbial community becomes unbalanced, allowing invading diseases to thrive. Treatments that include pre and probiotics, on the other hand, should be recommended. As a result, greater study into the use of probiotic therapy in the treatment of infectious diseases is needed. Indeed, our knowledge of the biology of complex diseases is expanding at an unprecedented rate and with unprecedented resolution, even as we recognize that what we have seen thus far is simply the tip of the iceberg, and that a large corpus of knowledge appears to be just around the corner. This should eventually lead to more effective treatments and prevention methods based on logical microbiota-based therapeutics. Therefore, researchers are using high throughput sequencing technologies and analytical methods, substantial advances have been achieved in both identifying the microbial taxa and understanding the relationship between microbiome composition and host phenotype, providing mechanistic insight on which microbes may be beneficial or which may be detrimental to one's health and given each microbiome is unique to an individual, this represents how high throughput sequencing technologies is impacting the future of personalized medicine and animal health, enhanced crop yield and nutritional quality, and the control of various pests and disease agents. In similar way, a complex microbiome minimizes the risk of some diseases, and probiotics can help with symptoms like IBS and eczema. Current tools and understanding of the microbiome have enabled researchers to develop new strategies to leverage applications of the microbiome. Overall, the grand vision of applied microbiome research is to improve health of humans, animals, plants, and whole ecosystem.

12. Future prospectus of microbiome

Microbiome research generates a large quantity of data, which necessitates the use of advanced computational techniques, which are rapidly evolving. Furthermore, many of the existing mathematical tools analyze connection rather than causation. As a result, researchers should remember that microbiome characterization, data analysis, and modelling are only a small part of the discovery process, and that they should be used in conjunction with traditional in vitro and in vivo model studies to prove cause and effect. To advance microbiome research into the therapeutic domain, researchers must go beyond clinical association studies to validate their models in other clinical cohorts and understand the mechanisms of causation in vitro, ex vivo, and animal model systems. While metagenomic studies have revealed immense diversity, additional tools are required to understand the community structure, function, and their interaction with host environments. Using microbiome analysis with Next generation sequencing to help define biomarkers and stratify patient populations, which may help improve therapeutic outcomes in the future. In addition, gaining deeper understanding of the microbiome through improved tools and methods will enable engineers and innovators to develop better applications and unlock the potential of the microbiome. As a result, pre/probiotics are likely to be coupled with other dietary substances to generate a more powerful health benefit. Furthermore, merging multiple study disciplines and employing new technological approaches in microbiome research is predicted to open the way for the development of evidence-based clinical therapies for modern-day health challenges. Previous research has found that bacteria have 35.5 million functions, of which just 0.02% are known, according to computational predictions. Despite the growing body of research on the microbiome, our understanding of its function, particularly how it influences health and disease, is limited because to the lack of a "universal" standard for study comparability.

The microbiome has been the subject of significant attempts to unravel all of its genetic information for the benefit of humankind, which requires worldwide coordination. Ultimately, new clinical tools and applications should be developed, as each individual and every population requires more individualized and effective care.

Conflict of interest

The authors declare that they have no conflict of interest.

Author details

Basavaraju M.*, Gunashree B.S. and Srinath B.S.
Department of Studies and Research in Microbiology, Mangalore University, Kodagu, India

*Address all correspondence to: microbasava@gmail.com

References

[1] Franzosa EA, Huang K, Meadow JF, Gevers D, Lemon KP, Bohannan BJM, et al. Identifying personal microbiomes using metagenomic codes. Proceedings of the National Academy of Sciences of the United States of America. 2015;**112**:2930-2938. DOI: 10.1073/pnas.1423854112

[2] Dekaboruah E, Suryavanshi MV, Chettri D, Verma AK. Human microbiome: An academic update on human body site specific surveillance and its possible role. Archives of Microbiology. 2020;**202**:2147-2167

[3] Lederberg J, McCray AT. 'Ome sweet 'omics—A genealogical treasury of words. Scientist. 2001;**15**:8

[4] Whitman WB, Coleman DC, Wiebe WJ. Prokaryotes: The unseen majority. Proceedings of the National Academy Science USA. 1998;**95**(12):6578-6583

[5] Gill SR, Pop M, Deboy RT, Eckburg PB, Turnbaugh PJ, Samuel BS, et al. Metagenomic analysis of the human distal gut microbiome. Science. 2006;**312**(5778):1355-1359

[6] Roberfroid MB, Bornet F, Bouley C, Cummings JH. Colonic microflora: Nutrition and health. Summary and conclusions of an International Life Sciences Institute (ILSI) [Europe] workshop held in Barcelona, Spain. Nutritional Review. 1995;**53**(5):127-130

[7] Cash HL, Whitham CV, Behrendt CL, Hooper LV. Symbiotic bacteria direct expression of an intestinal bactericidal lectin. Science. 2006;**313**(5790):1126-1130

[8] Yadav M, Verma MK, Chauhan NS. A review of metabolic potential of human gut microbiome in human nutrition. Archives of Microbiology. 2018;**200**:203-217. DOI: 10.1007/s00203-017-1459

[9] Human Genome 10th Anniversary: Digging Deep into the Microbiome: https://www.sciencemag.org/content/331/6020/1008.summary

[10] Thaiss CA, Zmora N, Levy M, Elinav E. The microbiome and innate immunity. Nature. 2016;**535**(7610):65-74

[11] Belkaid Y, Hand TW. Role of the microbiota in immunity and inflammation. Cell. 2014;**157**(1):121-141. DOI: 10.1016/j.cell.2014.03.011

[12] Walaa KM, Fadia C, Shannon H. Recent advances in understanding the structure and function of the human microbiome. Frontiers in Microbiology. 2022;**13**:825338. DOI: 10.3389/fmicb.2022.825338

[13] David A. The GUT Microbiome: New Understanding and Potential Translational Applications for Disease Management. UK: Nova Science Pub Inc; 2015

[14] Glendinning L, Stewart RD, Pallen MJ, et al. Assembly of hundreds of novel bacterial genomes from the chicken caecum. Genome Biology. 2020;**21**:34

[15] De Sordi L, Lourenço M, Debarbieux L. The battle within: Interactions of bacteriophages and bacteria in the gastrointestinal tract. Cell Host & Microbe. 2019;**25**(2):210-218. DOI: 10.1016/j.chom.2019.01.018

[16] Henry LP, Bruijning M, Forsberg SKG, et al. The microbiome

extends host evolutionary potential. Nature Communication. 2021;**21**:5141

[17] Cullen CM, Aneja KK, Beyhan S, Cho CE, Woloszynek S, Convertino M, et al. Emerging priorities for microbiome research. Frontiers in Microbiology. 2020;**11**:136. DOI: 10.3389/fmicb.2020.00136

[18] Ley RE, Hamady M, Lozupone C, Turnbaugh PJ, Ramey RR, Bircher JS, et al. Evolution of mammals and their gut microbes. Science. 2008;**320**:1647-1651. DOI: 10.1126/science.1155725

[19] Moeller AH, Suzuki TA, Lin D, Lacey EA, Wasser SK, Nachman MW. Dispersal limitation promotes the diversification of the mammalian gut microbiota. Proceedings of the National Academy Science USA. 2017;**114**:13768. DOI: 10.1073/pnas.1700122114

[20] Grice EA, Segre JA. The human microbiome: Our second genome. Annual Review of Genomics and Human Genetics. 2012;**13**:151-170

[21] https://www.thegutmicrobiome.com/factors-that-influence-gut-microbiota

[22] Lloyd PJ, Abu AG, Huttenhower C. The healthy human microbiome. Genome Medicine. 2016;**8**:51

[23] Koppel N, Maini RV, Balskus E. Chemical transformation of xenobiotics by the human gut microbiota. Science. 2017;**2017**:356

[24] Johnson CL, Versalovic J. The human microbiome and its potential importance to pediatrics. Pediatrics. 2012;**129**(5):950-960. DOI: 10.1542/peds.2011-2736

[25] Den Besten G, Van Eunen K, Groen AK, Venema K, Reijngoud DJ, Bakker BM. The role of short-chain fatty acids in the interplay between diet, gut microbiota, and host energy metabolism. Journal of Lipid Research. 2013;**54**(9):2325-2340. DOI: 10.1194/jlr.R036012

[26] Morowitz MJ, Carlisle E, Alverdy JC. Contributions of intestinal bacteria to nutrition and metabolism in the critically ill. The Surgical Clinics of North America. 2011;**91**(4):771-785

[27] Arumugam M et al. Enterotypes of the human gut microbiome. Nature. 2011;**12**(7346):174-180

[28] Canny GO, McCormick BA. Bacteria in the intestine, helpful residents or enemies from within. Infection and Immunity. 2008;**76**(8):3360-3373

[29] Jandhyala SM. Role of the normal gut microbiota. World Journal of Gastroenterology. 2015;**21**(29):8787-8803

[30] Gao QY, Chen YX, Fang JY. 2019 Novel coronavirus infection and gastrointestinal tract. Journal of Digestive Diseases. 2020;**21**:125-126. DOI: 10.1111/1751-2980.12851

[31] https://www.hmpdacc.org/overview/

[32] Galloway PJ, Hanson B. Tools for analysis of the microbiome. Digestive Diseases and Sciences. 2020;**65**(3):674-685. DOI: 10.1007/s10620-020-06091-y

[33] Kinross JM, von Roon AC, Holmes E, Darzi A, Nicholson JK. The human gut microbiome: Implications for future health care. Current Gastroenterology Report. 2008;**10**(4):396-403

[34] Lim YW, Schmieder R, Haynes M, Willner D, Furlan M, Youle M, et al. Metagenomics and metatranscriptomics: Windows on CF-associated viral and microbial communities. Journal of Cystic Fibrosis. 2013;**12**(2):154-164

[35] Culligan EP, Sleator RD, Marchesi JR, Hill C. Metagenomics and novel gene discovery: Promise and potential for novel therapeutics. Virulence. 2014;**5**(3):399-412. DOI: 10.4161/viru.27208

[36] Mousa WK, Chehadeh F, Husband S. Recent advances in understanding the structure and function of the human microbiome. Frontiers in Microbiology. 2022;**13**:825338

[37] Guo H, Chou WC, Lai Y, Liang K, Tam JW, June B. Multi-omics analyses of radiation survivors identify radio protective microbes and metabolites. Science. 2020;**370**:9097. DOI: 10.1126/science.aay9097

[38] Sani G, Manchia M, Simonetti A, Janiri D, Paribello P, Pinna F, et al. The role of gut microbiota in the high-risk construct of severe mental disorders: A mini review. Frontiers in Psychology. 2021;**11**:585-769. DOI: 10.3389/fpsyt.2020.585769

[39] Berg G. Plant-microbe interactions promoting plant growth and health: Perspectives for controlled use of microorganisms in agriculture. Applied Microbiology and Biotechnology. 2009;**84**:11-18

[40] Wang DD, Nguyen LH, Li Y, Yan Y, Ma W, Rinott E, et al. The gut microbiome modulates the protective association between a Mediterranean diet and cardio metabolic disease risk. Nature Medicine. 2021b;**27**:333-343. DOI: 10.1038/s41591-020-01223-3

[41] Goyal D, Ali SA, Singh RK. Emerging role of gut microbiota in modulation of neuro-inflammation and neurodegeneration with emphasis on Alzheimer's disease. Progress in Neuro-Psychopharmacology & Biological Psychiatry. 2021;**106**:110-112. DOI: 10.1016/j.pnpbp.2020.110112

[42] Das NK, Schwartz AJ, Barthel G, Inohara N, Liu Q, Sankar A, et al. Microbial metabolite signaling is required for systemic iron homeostasis. Cell Metabolism. 2020;**31**:115

[43] Barbaro NR, Foss JD, Kryshtal DO, Tsyba N, Kumaresan S, Xiao L, et al. Dendritic cell amiloride-sensitive channels mediate sodium-induced inflammation and hypertension. Cell Reports. 2017;**21**:1009-1020. DOI: 10.1016/j.celrep.2017.10.002

[44] Wibowo MC, Yang Z, Borry M, Hübner A, Huang KD, Tierney BT, et al. Reconstruction of ancient microbial genomes from the human gut. Nature. 2021;**594**:234-239. DOI: 10.1038/s41586-021-03532-0

[45] Faust K, Fah Sathirapongsasuti J, Izard J, Segata N, Gevers D, Raes J, et al. Microbial co-occurrence relationships in the human microbiome. PLoS Computational Biology. 2012;**8**:e1002606. DOI: 10.1371/journal.pcbi.1002606

[46] Carlson AL, Kai XM, Azcarate PA, Rosin SP, Fine JP, Wancen M, et al. Infant gut microbiome composition is associated with non-social fear behavior in a pilot study. Nature Communication. 2021;**12**:3294. DOI: 10.1038/s41467-021-23281-y

[47] Jin X, Zhang Y, Celniker SE, Xia Y, Mao JH, Snijders AM, et al. Gut microbiome partially mediates and coordinates the effects of genetics on anxiety-like behavior in collaborative cross mice. Scientific Reports. 2021;**11**:270. DOI: 10.1038/s41598-020-79538-x

[48] He Y, Kosciolek T, Tang J, Zhou Y, Li Z, Ma X, et al. Gut microbiome and magnetic resonance spectroscopy study of subjects at ultra-high risk for psychosis may support the membrane hypothesis.

European Psychiatry. 2018a;**53**:37-45. DOI: 10.1016/j.eurpsy.2018.05.011

[49] Maifeld A, Bartolomaeus H, Lober U, Avery EG, Steckhan N, Markó L, et al. Fasting alters the gut microbiome reducing blood pressure and body weight in metabolic syndrome patients. Nature Communications. 2021;**12**:1970. DOI: 10.1038/s41467-021-22097-0

[50] Huttenhower C, Gevers D, Knight R, Abubucker S, Badger JH, Chinwalla AT, et al. Structure, function and diversity of the healthy human microbiome. Nature. 2012;**486**:207-214. DOI: 10.1038/nature11234

[51] Larsen OFA, Claassen E. The mechanistic link between health and gut microbiota diversity. Scientific Reports. 2018;**8**:6-10. DOI: 10.1038/s41598-018-20141-6

[52] Sung J, Kim S, Cabatbat JJT, Jang S, Jin Y-S, Jung GY, et al. Global metabolic interaction network of the human gut microbiota for context-specific community-scale analysis. Nature Communications. 2017;**8**:15393

[53] Wang M, Karlsson C, Olsson C, Adlerberth I, Wold AE, Strachan DP, et al. Reduced diversity in the early fecal microbiota of infants with atopic eczema. The Journal of Allergy and Clinical Immunology. 2007;**121**:129-134

[54] Abrahamsson TR, Jakobsson HE, Andersson AF, Björkstén B, Engstrand L, Jenmalm MC. Low gut microbiota diversity in early infancy precedes asthma at school age. Clinical and Experimental Allergy. 2014;**44**:842-850. DOI: 10.1111/cea.12253

[55] Karlsson F, Tremaroli V, Nielsen J, Bäckhed F. Assessing the human gut microbiota in metabolic diseases. Diabetes. 2013;**62**:3341-3349. DOI: 10.2337/db13-0844

[56] Bisgaard H, Li N, Bonnelykke K, Chawes BLK, Skov T, Paludan-Müller G, et al. Reduced diversity of the intestinal microbiota during infancy is associated with increased risk of allergic disease at school age. The Journal of Allergy and Clinical Immunology. 2011;**128**:646-652. DOI: 10.1016/j.jaci.2011.04.060

[57] Ferreira CM, Vieira AT, Vinolo MAR, Oliveira FA, Curi R, Martins F. The central role of the gut microbiota in chronic inflammatory diseases. Journal of Immunologic Research. 2014;**2014**:689492. DOI: 10.1155/2014/689492

[58] Kennedy PJ, Cryan JF, Dinan TG, Clarke G. Irritable bowel syndrome: A microbiome-gut-brain axis disorder? World Journal of Gastroenterology. 2014;**20**:14105. DOI: 10.3748/wjg.v20.i39.14105

[59] Lakhan SE, Kirchgessner A. Gut inflammation in chronic fatigue syndrome. Nutrition Metabolism. 2010;**7**:79

[60] Castellarin M, Warren RL, Freeman JD, Dreolini L, Krzywinski M, Strauss J, et al. *Fusobacterium nucleatum* infection is prevalent in human colorectal carcinoma. Genome Research. 2012;**22**:299-306. DOI: 10.1101/gr.126516.111

[61] Kau AL, Ahern PP, Griffin NW, Goodman AL, Gordon JI. Human nutrition, the gut microbiome and the immune system. Nature. 2011;**474**:327. DOI: 10.1038/nature10213

[62] Geuking MB, McCoy KD, Macpherson AJ. Metabolites from intestinal microbes shape Treg. Cell Research. 2013;**23**:1339-1340. DOI: 10.1038/cr.2013.125

[63] Tremaroli V, Bäckhed F. Functional interactions between the gut microbiota and host metabolism. Nature. 2012;**489**: 242-249. DOI: 10.1038/nature11552

[64] Medina M, Izquierdo E, Ennahar S, Sanz Y. Differential immunomodulatory properties of *Bifidobacterium logum* strains: Relevance to probiotic selection and clinical applications. Clinical and Experimental Immunology. 2007;**150**:531-538. DOI: 10.1111/j.1365-2249.2007.03522.x

[65] Gupta S, Allen-Vercoe E, Petrof EO. Fecal microbiota transplantation: In perspective. Therapeutic Advances in Gastroenterology. 2016;**9**:229-239. DOI: 10.1177/1756283X15607414

[66] Spanogiannopoulos P, Bess EN, Carmody RN, Turnbaugh PJ. The microbial pharmacists within us: A metagenomic view of xenobiotic metabolism. Nature Reviews. Microbiology. 2016;**14**:273-287. DOI: 10.1038/nrmicro.2016.17

[67] Srinath BS, Shastry RP, Kumar SB. Role of gut-lung microbiome crosstalk in COVID-19. Research Biomedical Engineering. 2022;**38**(1):181-191. DOI: 10.1007/s42600-020-00113-4

[68] Baohong W, Mingfei Y, Longxian L, Zongxin L, Lanjuan L. The human microbiota in health and disease. Engineering. 2017;**3-1**:71-82

[69] Whiteside SA, Razvi H, Dave S, Reid G, Burton JP. The microbiome of the urinary tract—A role beyond infection. Nature Reviews. Urology. 2015;**12**:81-89

[70] Pascal M, Perez-Gordo M, Caballero T, et al. Microbiome and allergic diseases. Frontiers in Immunology. 2018;**9**:1584

[71] https://www.bbc.com/future/article/20211115-how-your-microbiome-can-improve-your-health

[72] Goodman's Medical Cell Biology. 4th ed. Elsevier. 2021. pp. 393-404

[73] Schirmer M et al. Linking the human gut microbiome to inflammatory cytokine production capacity. Cell. 2016;**167**:1897

[74] Geva-Zatorsky N et al. Mining the human gut microbiota for immunomodulatory organisms. Cell. 2017;**168**:928-943

[75] Vatanen T. Variation in microbiome LPS immunogenicity contributes to autoimmunity in humans. Cell. 2016;**165**:842-853

[76] Dubin K et al. Intestinal microbiome analyses identify melanoma patients at risk for checkpoint-blockade-induced colitis. Nature Communications. 2016;**7**:10391

[77] Mueller NT et al. Does vaginal delivery mitigate or strengthen the intergenerational association of overweight and obesity? Findings from the Boston Birth Cohort. International Journal of Obesity. 2005;**41**:497

[78] Gilbert JA, Blaser MJ, Caporaso JG, Jansson JK, Lynch SV, Knight R. Current understanding of the human microbiome. Nature Medicine. 2018;**24**(4):392-400. DOI: 10.1038/nm.4517

[79] Sommer MO. Advancing gut microbiome research using cultivation. Current Opinion in Microbiology. 2015;**27**:127-132

[80] https://www.bioaster.org/technology-innovations-in-microbiology-infectiology/fields-of-applications/microbiota/

Section 2

Microbiota and Diseases

Chapter 2

Gut Microbiota and Inflammatory Bowel Disease

Bahareh Vakili, Parisa Shoaei, Zahra Esfandiari and Seyed Davar Siadat

Abstract

Inflammatory bowel disease (IBD) is a chronic and relapsing inflammatory disorder that includes Crohn's disease and ulcerative colitis. Ulcerative colitis involves the distal colon, proximal colon, and cecum and can lead to ulcerations and bleeding. Crohn's disease appears as patched lesions in the gastrointestinal tract and inflammation, stenosis, or fistulas. IBD affects millions of people worldwide and has been associated with high morbidity and mortality. Our intestine is colonized by trillions of microorganisms (including bacteria, viruses, fungi, and protozoa), which constitutes the microbiota. Reduction of bacteria with anti-inflammatory capacities and increase of bacteria with inflammatory capacities are observed in patients with IBD when compared with healthy individuals. Microbial balance is needed for the development of a healthy gut and a symbiotic microbiota without problems. Any disturbance in that balance leads to dysbiosis and the host may become more susceptible to disease. Some alteration in the microbiome is protective or causative; thus, we selectively will review IBD disease, pathogenesis, and potential roles of some members of microbiota in IBD. In this chapter, we also explain the therapeutic approaches targeting microbiota (probiotics, prebiotics, postbiotics) and the relationship between gut microbiota imbalance, and how defects in this dysbiosis can lead to disease.

Keywords: inflammatory bowel disease, gut microbiome, Crohn's disease, ulcerative colitis, microbiome, dysbiosis, therapy

1. Introduction

Inflammatory bowel disease (IBD), including ulcerative colitis (UC), Crohn's disease (CD), and indeterminate colitis, is a chronic and relapsing inflammatory disorder of the gastrointestinal tract [1, 2]. More than 1 million residents in the United States and 2.5 million individuals living in Europe are estimated to be suffering from IBD [2]. The incidence of IBD has been rapidly increasing in newly industrialized countries in Asia, the Middle East, Africa, and South America over the last two decades [3, 4]. IBD has been associated with high morbidity and mortality, low quality of life, and financially demanding medical care [5]. The causes of this disease are multifactorial, the two main types: UC and CD, have similar clinical and

IntechOpen

pathological presentations and can cause irreversible impairment of the structure and function of the gastrointestinal tract [6]. These diseases are characterized by a relapsing behavior, manifested by alternating phases of inactive states in which there is no intestinal inflammation and active states that present inflammation or any other disease symptoms [7]. Although the main biological processes involved in the development of both conditions are different [6]. CD can affect any part of the GI tract, especially the terminal ileum, associated with inflammation, stenosis, and/or fistulas [7, 8].

CD occurs in patients between the ages of 15 and 35 years, affects the mucosal layer of the colon, and causes abdominal pain, diarrhea, and fever, fistula, lesions in the rectum or intestine, and other symptoms. CD damages the small intestine; therefore, malnutrition is very common in CD [9]. Despite the UC, rectal bleeding is less common in CD patients and more than 50% of patients with CD suffer from folate and vitamin D deficiency, while more than 50% of people with UC suffer from iron deficiency [10].

UC disease is a mucosal inflammation that can only affect the large intestine, i.e., the colon, and the inflammation generally starts in the distal colon, going forward through the proximal colon until the cecum and can lead to ulcerations and bleeding [11]. About 25% of UC patients are diagnosed before the age of 18 years, because this disease affects adolescence [9]. There are different diagnostic tests for UC including clinical, endoscopic, histologic, and radiological tests although approximately 8.5% of IBDs are unclear [12, 13]. UC is the initial subtype of IBD, and the term IBD includes the characteristics of both CD and UC. It has long been difficult to distinguish between these two diseases, but now there is a clinical definition for both. Both diseases can affect specific parts of the lives of patients, such as school, job, social life, and family life **Figure 1** [9].

The concept of IBD pathogenesis is based on the theory of a disrupted intestinal barrier and a dysregulated immune response in a genetically susceptible host. IBD presents defects in the detection and control of the gut microbiota, associated with unbalanced immune reactions, genetic mutations that confer susceptibility to the disease, and complex environmental conditions such as a Westernized lifestyle [7, 14]. There is a strong clustering in families and with certain ethnicities. Other studies showed 15–50 times increased relative risk for siblings of a CD patient to also develop CD. The ethiopathology of IBD is multifactorial and is characterized by the interaction between genetic, microbial, environmental, and life style factors, which influences the immune responses and leads to the gut inflammation. Gut microbiota is important for the development and maturation of the immune system and reduced microbial diversity and its dysbiosis observed in IBD patients (**Figure 1**).

More than 200 IBD-associated susceptible genes have been identified, some of which are known to be involved or implicated in mediating host responses to gut microbiota [14]. This has evoked the possibility that gut microbiota is implicated in the pathogenesis of IBD [3]. Microbial factors have been historically proven to be indispensable for the onset of IBD, and advances in high-throughput sequencing have enabled us to elucidate the gut microbiome in IBD. IBD can be caused by determined infection of an enteric pathogen such as *Mycobacterium avium* subspecies paratuberculosis, *Clostridioides difficile, Helicobacter pylori*, *Campylobacter concisus*, *Fusobacterium nucleatum,* and adhesion-invasive *Escherichia coli*. An excess of translocation of intestinal bacteria across the intestinal barrier, the imbalance between beneficial and detrimental commensal bacteria can cause IBD [15, 16]. Some bacteria

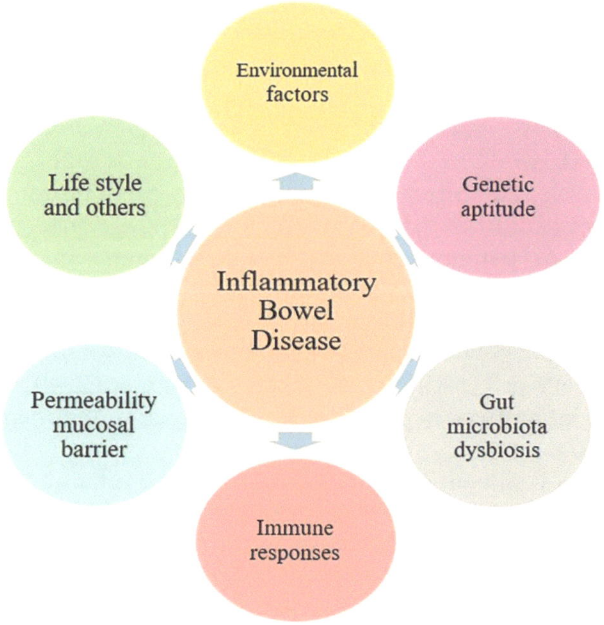

Figure 1.
The effective agents in development of inflammatory bowel disease.

including AIEC can be considered as both a persistent pathogen and detrimental commensal bacteria [17].

Therefore, this chapter covers the origins, causes, diagnosis, and treatment strategies of this complex disease.

2. IBD and immunity system

Epithelial layer integration permits the gastrointestinal bacteria to communicate with the immune system [18]. The mucosal layer is the first physical barrier on the mucosal surface and is produced by the polymerization of gel-forming mucins secreted by Goblet cells. The second defense barrier against bacterial attack is the intestinal epithelium, which makes up of enterocytes and particular epithelial cells called Goblet and Paneth cells [19]. Intestinal epithelial cells prevent the influx of antigens and the attack of pathogens and commensal microbes [18]. Intestinal epithelial cells (IECs) also express toll-like receptors (TLRs) and nucleotide oligomerization domain receptors (NODs), which are pathogen-sensitive innate immune receptors. IECs then make chemokines and cytokines to engage immune cells [18]. TLR signaling pathways helps the epithelial barrier to remain intact and produce 12 and interleukin 6 [18, 20]. The epithelial barrier impairment causes intestinal permeability to increase, which has been shown in CD and also in UC, and this might be a main pathogenetic mechanism in IBD [19]. TLR acts as pro/anti-inflammatory gene activation inducer and controls the adaptive immune responses [21, 22].

Intestinal immune cells including innate immune cells and adaptive immune cells significantly involve in immune responses in IBD [23]. Macrophages, TLRs, and NOD-like receptors (NLRs) are essential for developing tolerance to certain pathogens and promoting wound treatment. Binding to pathogene receptors leads to the activation of different signaling pathways and the production of proinflammatory cytokines, chemokines, and antimicrobial peptides. The antigen-presenting cells (APCs) link innate immunity and adaptive immunity by secreting cytokines and presenting antigens to the T cells [24]. Fine gut-resident macrophages, described by a lack of CD14 expression, manifest decreased response, proliferation, and chemotactic activity. The gut-resident macrophages have increased phagocytic activity and secretion of cytokines in IBD patients, causing dramatic inflammation [25]. After microorganisms' invasion, innate immunity activates after a few hours [26]. Macrophage cells kill specific pathogens, such as peptides and lipopolysaccharides. In IBD acute phase, the number of macrophages in the intestinal mucosa increases dramatically, and a large number of T cells and costimulatory molecules such as CD40, CD80, and CD86 are involved in the inflammatory process and intolerance of commensal microbes and immune activity [27].

Malfunction in TLR signaling can induce an intestinal inflammatory response with various clinical phenotypes, including the IBD. A considerable target of the TLR signaling is the activation of the transcription factor NF-kB, which regulates the expression of a variety of genes responsible for controlling the innate response, such as IL-1, IL-2, IL-6, IL-12, and TNF-α [28, 29]. **Table 1** shows the cytokines and cellular sources involved in immune response in IBD. Both IL-1 and TNF-α share numerous pro-inflammatory properties responsible for the development of IBD [30]. Dendritic cells are professional antigen-presenting cells that activate T cells and induce adaptive immune responses, describing key players in the cross talk between innate and adaptive immunity [38].

The other IBD risk variants in other genes are involved in IL-12 and *CCR6*, chemokine receptors preferentially expressed on IL-17 producing cells [19]. IL-23/IL- 17 axis has a key role in this cross talk and the *IL23R* gene encodes a specific subunit of the IL23 receptor that has been identified and largely replicated in independent cohorts of patients with both CD and UC [45]. Other clinical studies have found that the intestinal mucosa and lamina propria of IBD patients contain much higher levels of Th17 cells, IL-17, and IL-23 compared with the healthy controls [24].

Appositive of the innate immune response, the adaptive immune system is very specific, it presents long-lasting immunity. Key players of the adaptive immune response are T cells. Th0 cells can become activated and either differentiate into Th1 or Th2 or Th17 cells [19, 38]. However, a dysregulated T cell response with abnormal development of activated T cell subsets causes inflammation because of an excess release of cytokines and chemokines, which have multiple pathogenic impacts on components of the immune system. **Figure 2** shows the immune response in IBD. The levels of T-cell-derived cytokines detected in IBD mucosa, different studies have associated CD and UC with different subtypes of pro-inflammatory immune responses. Therefore, the innate immune response is as important as the adaptive immune system in inducing gut inflammation in these patients [19, 24, 38].

Genome-wide association studies and immunological studies have mentioned that IBD pathogenesis is related to mucosal innate immune responses, including classical Th1 response in CD patients and Th2 type-like response in UC patients [45, 46].

In mouse model studies, induction of CD caused increase of IFN-γ expression in their spleen and local intestinal mucosa [43]. CD evolution is generally mediated

Pro-inflammatory			
Cytokines	**Cellular Sources**	**Principle function**	**Role in immune system**
IL-1, IL-1β	Mϕ, IECS, Monocytes	Influence on the T cell and secretory cytokines	Innate immune response [30–32]
IL-2	Th-cells	T & B cells proliferation IFN-γ production	Adaptive immune response [31]
IL-6	DCs, Mϕ	Differention of Th17, Treg cells and activating STAT-3 signaling pathway	Innate immune response [31, 33–35]
IL-12	DCs, Mϕ	Promoting the differentiation of Th1 and Th17 cells	Innate immune response [36, 37]
IL-13	Th2 cells	Intestinal permeability inducing and activating of B cell	Innate immune response [38]
IL-17	Th17 cells	Inducing and promoting of secretory cytokines	Adaptive immune response [39, 40]
IL-18	Mϕ	Provoking the secretion of pro-inflammatory cytokines	Innate & adaptive immune response [32]
IL-22	Th17 cells	Inhibiting pathogens of intestinal and repairing of intestinal tissue	Innate immune response [41]
IL-23	Mϕ	Provoking the production cytokines	Innate immune response [36]
TNF-α	Mϕ, DCs, Th-cells	Promoting the production cytokines and Th-cells proliferation	Innate immune response [31, 42]
INF-γ	Th-cells	Activating NF-κB signaling pathway and activating of Mϕ	Adaptive immune response [43]
Anti-inflammatory			
IL-4	Th2-cells	Th2-cells differentiation and inhibiting the production of cytokines of Th1cells	Adaptive immune response [31]
IL-10	DCs, Mϕ, Treg cells	Inhibiting the production of cytokines Th1 cells	Adaptive immune response [31, 44]
TGF-β	DCs, Treg cells, T cells	Treg and Th17 differentiation and restraining of Th-cells	Adaptive immune response [31, 34, 35]

Abbreviations: Mϕ: Macrophage, IECS: Intestinal epithelial cells, DCs: dendritic cells, STAT-3: signal transducer and activator of transcription, NF-κB: nuclear factor kappa B.

Table 1.
The pro-inflammatory agents' contribution in immune response in IBD.

by CD4+ Th1 and Th17 cells, and IFN-γ is a major cytokine declared in this disease [47]. Deficiencies of IL17-A and IL17-B in experimental models showed both pro-inflammatory and tissue-protective effects against colitis depending on the model used [19, 48]. However in mucosa of IBD patients, IL-17A cells regulate and induce a number of pro-inflammatory molecules [38].

Regulatory T cells (Treg) produce the anti-inflammatory cytokines (IL-10, TGF) and exert an effective anti-inflammatory action in experimental colitis. Treg are reduced in peripheral blood of patients with active IBD in comparison with quiescent IBD patients and control subjects [49, 50]. In contrast, Treg are increased

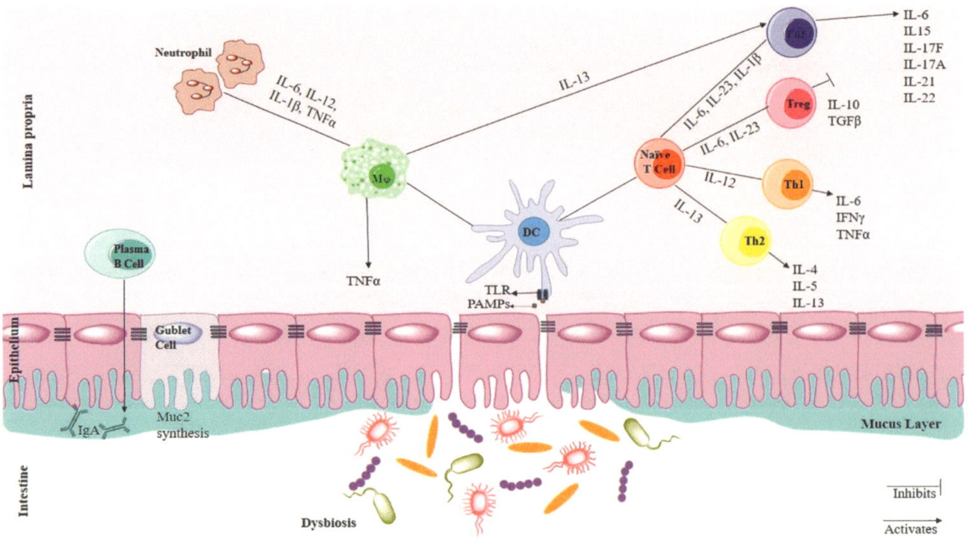

Figure 2.
The role of immune response on progress IBD.

in the intestinal mucosa of IBD patients, and their function is normal. An intact TGF signaling, which is impaired in inflamed IBD mucosa because of upregulation of the inhibitory molecule Smad7, is needed for Treg function [19]. Treg cells, expressing the transcription factor forkhead box P3 (FOXP3), have a negative immunomodulatory character in immune tolerance and a crucial role in the pathogenesis of IBD [24, 51, 52].

3. IBD and gut microbiota

IBD is obviously related to gut dysbiosis that impairs host-microbe and immune homeostasis [53]. The human gut includes trillions of commensal bacteria per gram of gut lumen content. These bacteria can be nutritious and provide the intestinal epithelium [38, 54]. The gut microbiota leads to intestinal homeostasis due to our physiological procedure and metabolites [55]. There are different phyla, including Bacteroidetes, Firmicutes, Proteobacteria (*Escherichia* and *Helicobacter*), and Actinobacteria that include fungi, protists, and viruses (**Table 2**) [56, 57].

Ecological factors, such as as host diet, hygiene, antibiotic consumption, and lifestyle, induce immune responses that change the intestinal microbiota and damage the mucosal barrier [38, 58]. Gut microbiota plays an important role in the pathogenesis of IBD and impacts energy metabolism host, immune homeostasis, development and maintenance of mucosal integrity [24]. **Table 3** shows the effect of gut microbiota in inflammatory bowel disease and its interdependence with the immune response.

For example, *Clostridium* cluster IV and XIVa were less abundant in IBD patients than in healthy controls [55]. Bacteroides genus is obligate anaerobe bacteria and consists a large amount of the normal gut microbiota. *B. fragilis* decreases in IBD patients and promotes the quantities of anti-inflammatory cytokines against colitis [24, 75]. The overgrowth of *Enterobacteriaceae, Pseudomonas*-like bacteria, and *Escherichia coli* promotes the intestinal inflammation and alters the composition of

Increased	Decreased	
Bacteria	*Fusobacterium* species	*Bififidobacterium* species
	Pasturellaceae Bacteroides species	
	Proteobacteria (adherent invasive *Escherichia coli*)	*Clostridium* XIVa, IV
	Ruminococcus gnavus Faecalibacterium prausnitzii	
	Veillonellaceae Roseburia species	
	Suterella species	
Fungi	*Candida albicans*	*Saccharomyces cerevisiae*
	Candida tropicalis	
	Clavispora lusitaniae	
	Cyberlindnera jadinii	
	Kluyveromyces marxianus	
Viruses	Caudivirales	

Table 2.
Microbiota changes associated with inflammatory bowel disease.

the microbiota in most colitis models and IBD patients [58, 76, 77]. *Faecalibacterium prausnitzii* secret anti-inflammatory cytokines, which reduce in the intestine of IBD patients [24, 55]. *Fusobacterium* and *Ruminococcus gnavus* have also been increased in CDI patients [78]. In a recent study performed on IBD patients, a functional gut microbiome dysbiosis and impaired microbial transcript were seen. Facultative anaerobes were raised at the expense of obligate anaerobes [79]. Other different studies showed that the diversity of gut microbiota was either decreased or equal in IBD patients versus controls. *F. prausnitzii*, *Eubacterium rectale,* and *Akkermansia* were decreased, and *Actinomyces*, *Veillonella,* and *E. coli* were increased in patients with UC (**Table 3**) [80].

Other possible pathogens in the exacerbation of the IBD disease are *Mycobacterium avium* subspecies *paratuberculosis*, *Clostridium difficile*, *Listeria monocytogenes*, and Campylobacter *concisus*, as well as viruses, including *cytomegalovirus, Epstein-Barr virus,* and measles virus [17, 81]. In addition, a number of pathogenic parasites may involve in the progression of this disease. Overexposure of immune system in the presence of too many bacterial materials could also cause the loss of immunological tolerance to the bacteria, which are generally considered the normal flora in the gut [81]. Some of the individual bacterial species that associate with human IBD are reviewed here.

3.1 Clostridioides difficile

C. difficile is an obligate anaerobic Gram-positive spore-forming bacterium, which is prevalent in nature and also colonizes the human intestinal tract [81, 82]. *C. difficile* leads to diarrhea and colitis, frequently in persons who have been treated with antibiotics for other medical complications [81, 83, 84].

C. difficile can produce toxins type A and B, and IBD patients with *C. difficile* infections (CDI) appear severe clinical symptoms, such as abdominal pain, diarrhea, bloody stools, and leukocytosis [85, 86].

Depleted	Immune Association
SCFA producing bacteria (*F. prausnitzii, Roseburia, Eubacterium*)	Produce SCFA plays a major role in modulation of inflammation, regulation of immune responses, maintenance of barrier integrity in the gut, enhanced expansion of the Treg population, and skew of human dendritic cells to prime IL-10-secreting T cells [59–61].
Bacteroides fragilis	Produces lipid antigens controlling homeostatic iNKT cell proliferation and activation [62].
Bifidobacterium	Inhibits intestinal inflammation by acting on Treg cells [63].
Mbb. smithii	Weak association with pro-inflammatory mechanisms [64].
Enriched	**Immune association**
E. coli (adherent invasive)	Invades intestinal epithelial cells replicate in macrophages and induce granulomas [65].
Clostridiaceae (class) Clostridiales	*Clostridium difficile* could induce the expansion of regulatory Tcells (Treg) and to mitigate intestinal inflammation [66].
Proteobacteria (Salmonella, Yersinia, Desulfovibrio, Vibrio Helicobacte)	Associated with a pro-inflammatory state as revealed by quantification of common pro-inflammatory interleukins. The inflamed gut appears to provide a favorable environment for the expansion of this phylum [67].
R. gnavus	Secretes a complex glucomannan polysaccharide inducing TNFα secretion by dendritic cells [68].
Fusobacterium	Especially *F. nucleatum*, which is a well-recognized proinflammatory bacterium and it may secrete Outer Membrane Vesicles (OMVs) activate epithelial TLR4 to drive inflammation [69, 70].
C. albicans	Interacts with mucosal innate immune cells through the pathways associated with Dectin-1 in macrophages [71].
Bacteriophages (Caudovirales and Microviridae)	role in physiology of intestinal or change the bacterial in gut microbiota via predator-prey relationships [72]. Enterobacteria are the hosts of Microviridae [73].
Eukaryotic viruses	Infect of intestinal and develop host susceptibility to IBD by immune response via inflammatory mediators, and inducing alterations in the composition of the commensal bacteria [74].
Eukaryotic viruses	Infect host cells may increase host susceptibility to IBD by supporting a long-standing immune response through inflammatory mediators, as well as by inducing alterations in the composition of the commensal microbiota [74]
M. stastmanae	This leads to the substantial release of proinflammatory cytokines in monocyte-derived dendritic cells [64].

Table 3.
Gut microbiome in inflammatory bowel disease and its associations with the immune system [7].

CDI causes relapsed IBD, and IBD patients in remission had a significantly higher presence of toxigenic *C. difficile* in their intestinal tract as compared with healthy controls [87]. UC patients have a high risk of CDI in comparison with healthy population or CD patients.

The two toxins encoded by *tcdA* and *tcdB* genes lead to the disruption of epithelial cytoskeleton and tight junctions, which contribute to the CDI [81, 88, 89]. A reduction in butyrate-producing bacteria and increase in lactic-acid-producing bacteria were seen in CDI status. Overrepresentation of *Akkermansia* may be a predictive marker for the development of nosocomial diarrhea, which can result in a worse CDI prognosis [82]. Activation of the production of multiple inflammatory cytokines such as IL-8, TNF-a, IL-1, and tumor necrosis factor (TNF-a) could damage the intestinal epithelial cells and trigger IBD in CDI patients [90]. Reduced bile salts happen in the colon of patients with IBD leading to spore germination of C. difficile [86, 91]. Patients with IBD present common infections such as gastrointestinal infections of *C. difficile, Salmonella, Shigella,* and *Campylobacter jejuni* [81, 92].

3.2 *M. avium* subspecies *paratuberculosis*

M. avium species is commonly present in the environment and comprises four subspecies, including *M. avium* subspecies *avium, M. avium, M. avium* subspecies *hominissuis,* and *M. avium* subspecies *silvaticum* [93]. *M.avium* causes production of some inflammatory cytokines in IBD patients [86]. In IBD patients, the increase in metalloprotease leads to dysregulation in immune system and large level of inflammatory cytokines [94, 95]. Combination of multiple antibiotics including rifabutin, clofazimine, and clarithromycin, adds up to ciprofloxacin, metronidazole, or ethambutol, which are used for treatment of patients with positive different species of *M.avium* [96]. Also antibiotics such as nitroimidazoles and clofazimine are effective in the treatment of CD [97].

3.3 *Helicobacter pylori*

Helicobacter species are Gram-negative bacteria. H. pylori is an important pathogen that isolates from gastrointestinal tract of humans and animals. H. pylori infection has been reported in IBD patients and shows a protective effect in IBD [98, 99]. H. pylori increases the expression of forkhead box P3 (FOXP3) with stimulating of the regulatory T cells production, reduces the production of inflammatory cytokines, and finally, decreases inflammation [100, 101]. H. pylori with cytotoxin-associated gene A (CagA+) genotype, in IBD patients, diverts TH1 response to TH2 response that has anti-inflammatory task [102].

Helicobacter species are more detected in intestinal biopsies of patients with CD and UC than controls, although this difference was not significant [103]. Molecular studies detected non-pylori Helicobactor by Helicobacteriaceae family-specific PCR in 3% of IBD patients and 8% controls [104].

3.4 *C. concisus* and *Fusobacterium nucleatum*

Most strains of campylobacter colonize in the intestinal tract, but the colonization of *C. concisus* is in the oral cavity [105]. *C.concisus* is associated with IBD in the adult patients [106]. The virulence factors of *C. concisus* infect the lower parts of the intestinal tract [86]. Zonula occludens toxin (Zot) is expressed through a CON-Phi2 prophage and leads to the permeability of the epithelial cells and formation of IBD. This mechanism is similar to the *Vibrio cholerae* toxin [107]. *C. concisus* breaks the intestinal epithelial barrier and leads to apoptosis in human intestinal epithelial and intestinal inflammation [108]. The invasive strains of *C. concisus* enable them to survive in harsh conditions such as in anaerobic conditions [86].

F. nucleatum is an anaerobe bacterium that colonizes the oral cavity and intestinal tract [81]. It is abundant in intestinal tract of UC and IBD patients, and the quantity was linked with disease severity [109]. *F. nucleatum* leads to the damage of intestinal epithelium and promotes intestinal inflammation by inducing autophagic epithelial cell death [86].

3.5 Adherent-invasive *E. coli*

Adherent-invasive *E. coli* (AIEC) is a commensal human gut bacterium and is associated with ileal CD in the adult population. AIEC strains can adhere to and invade intestinal epithelial barrier assessed [110]. AIEC strains have various mechanisms and virulence factors, which are involved in the pathogenesis of IBD patients [86]. Several factors such as type 1 pili adhesion FimH and carcinoembryonic antigen cell adhesion molecule 6 are associated in promoting inflammation [111]. AIEC strains induced production of cytokines such as IL-8, TNF-a, and IL-6 in both epithelial cells and macrophages. Replication of AIEC in macrophages did not cause macrophage death, but increased production of TNF-a and IL-6 [81, 112].

4. Therapeutic approaches targeting microbiota (probiotics, prebiotics, postbiotics, and antibiotics)

Probiotics are live microorganisms, which allocate great health advantages for the host organism when used in an appropriate quantity [113]. Probiotics induce anti-inflammatory effects, enhance or renew barrier work, promote the growth of beneficial bacteria, and inhibit the growth of pathogens [114]. Probiotics rebalance the gut microflora shifting from pro- to anti-inflammatory state [115]. Prebiotics are substrates that are selectively utilized by probiotics allocating health benefits [17]. Inulin is a prebiotic that retains microbial population, helps the epithelium barrier function, and inhibits from pathogens translocation [116]. This process leads to the treatment of functional symptoms in IBD. Postbiotics are bioactive molecules produced by probiotics [117]. There are many reports that showed some probiotics and prebiotics can be beneficial in treatment and prevention of IBD in both human and mice models [118].

In CD, evidence for prebiotics and probiotics is commonly dissatisfactory and antibiotics have moderate effects [66]. The most common strains that are used as beneficial probiotics are *Bifidobacterium* species, *Enterococcus faecium*, *Lactobacillus* strains, and *Saccharomyces boulardii*, *Bacillus* species, and *Pediococcus* [115]. The theoretical risks of probiotics on animal models of IBD in different studies are described that include systemic infections, harmful metabolic activities, extreme immune stimulation in susceptible patients, gene transfer, and gastrointestinal adverse effects [119]. Some traditional probiotics, such as the probiotic cocktail VSL# 3 (containing a mix of four *lactobacilli*, three *Bifidobacteria,* and one *Streptococcus* strain), have shown limited effect in treating CD and UC, by reducing active inflammation and recurrence [17, 66].

Some clinical trials showed that *Lactobacillus rhamnosus* administration in gastroenteritis children did not have better outcomes than those who received placebo [120]. Although a multi-strain probiotic (including *L. rhamnosus*, *Lactobacillus plantarum*, *Lactobacillus acidophilus,* and *E. faecium*) is related with lower intestinal inflammation in UC patients, but not in CD patients [121]. Besides the mentioned traditional probiotics, *Akkermansia muciniphila* and their supernatants that contain postbiotics significantly reduced the severity of colitis [17]. *F. prausnitzii* produce barrier improving immunosuppressive SCFAs, stimulate Tregs to produce IL-10, which have protective effects on the intestine [66]. In mice models, administration of *A. muciniphila* or its postbiotic reduced the infiltrating macrophages and CD8+ in clolon and inhibited colitis [122, 123].

These helpful microbes and their metabolites should be investigated as therapeutic determinants in treatment of IBD. Dietary substrates such as oligosaccharides and fiber are prebiotics that selectively increase the quantities of SCFA-producing commensals, blocking the AIEC epithelial adherence, and the virulence products of intestinal pathogens in IBD [66].

Probiotic engineering with emerging technologies such as as CRISPR-Cas system can be used to produce to treat untreatable chronic inflammatory conditions [115]. With increasing our knowledge about viable bacterial strains and synthetic biology tools, we can identify and characterize extra probiotic bacterial strains as potential candidates for probiotic engineering [124].

Antimicrobial agents and IBD have a complex relationship. They have hazardous influences on the homeostasis of the host microbiota, leading to a population shift described by increased *Enterobacteriaceae* and decreased Clostridia abundant, which is regarded a possible pre-IBD condition [125]. Also, IBD patients treated with antibiotics are at high risk of forming an overgrowth of pathogenic microbes including *C. difficile*, candida, and bacteriophages [126].

In addition, antibiotics are an integral part of the treatment repertoire in IBD, whereas before the period of immunomodulation and biologic therapy. The mechanisms of antibiotics in treatment of IBD are a direct effect on the gut microbiota, preferring flora that are linked with anti-inflammatory properties, e.g., Bacteroides and Firmicutes, and decreasing pathogenic microbes that are associated with inflammation, such as as Enterobacteriaceae, e.g., E. coli and Fusobacterium [75]. Furthermore, we can choose target-specific pathobiants or to manage individual microbiome in IBD patients by determining patient stool samples prior to treatment [124].

The immunological mechanisms of IBD have made great upgrades, provided novel tactics for IBD treatment. Biological agents induce and maintain clinical remission of IBD and promote mucosal curing. A number of biological agents that have been approved for the treatment of IBD are some of the TNF-α inhibitors such as Infliximab, Adalimumab, Certolizumab pegol, Glimumab, Etanercept, and Tocilizumab [24]. However 10–40% of IBD patients do not respond or lose their response to treatment over time [127].

4.1 Fecal microbiota transplantation (FMT)

FMT appears effective therapy for treatment of recurrent CDI and in UC or CD remission induction but remains strong and safe in the long term is not clear [128, 129]. A significant proportion of recurrent CDI patients have IBD, and FMT is moderately less successful in treatment of CDI from patients with IBD in comparison with patients without IBD [130]. Some issues could affect the FMT outcome in IBD treatment including donor choice; preparation of fecal material; clinical management, the high abundances of fungi or virus communities in donor stool or other essential necessities for implementing an FMT center [131, 132].

Recently, the field of IBD genetics has made enormous progress, and different relative molecular and cellular pathways exist. Fluctuations in specific gene loci promise therapeutics for IBD in the future. Besides, FMT, novel natural medicines, new antimicrobial agents, and combined treatment programs are also anticipated to break the IBD and therapeutically delay. The combined treatment strategies that use anti-inflammatory agents and anti-fibrotic drugs will provide great insights into the existing IBD therapeutics [17].

5. Conclusion

Correct interplay between gut microbiota and the host is essential for human health. Microbial balance is pivotal for host metabolic and immune functions as well as to prevent disease development. Disturbance in that balance generates dysbiosis making the host susceptible to certain diseases. Gut microbiota stimulates the immune system, and altered composition of this microbiota in early life can lead to an inadequately trained immune system that can overreact to commensal microbes and lead to inflammatory diseases. Recent research has provided striking findings supporting that the gut microbiome plays an important function in the etiopathogenesis of IBD.

The clinical and epidemiological evidences showed that the infectious pathogens have possible role in IBD progression, especially, *Mycobacterium avium paratuberculosis*, *C. difficile*, *E. coli*, and *C. concisus*. Also, some viruses such as cytomegalovirus, Epstein-Barr virus, and measles by different pathogenesis have been be associated with the higher IBD risk; however, *H. pylori* may reduce intestinal inflammation and protect against IBD.

Author details

Bahareh Vakili[1], Parisa Shoaei[2*], Zahra Esfandiari[3] and Seyed Davar Siadat[4]

1 Isfahan University of Medical Sciences, Infectious Diseases, and Tropical Medicine Research Center, Isfahan, Iran

2 Isfahan University of Medical Sciences, Nosocomial Infection Research Center, Isfahan, Iran

3 Department of Food Science and Technology, Isfahan University of Medical Sciences, Food Security Research Center, School of Nutrition and Food Science, Isfahan, Iran

4 Department of Mycobacteriology and Pulmonary Research, Pasteur Institute of Iran, Tehran, Iran

*Address all correspondence to: shoaei@idrc.mui.ac.ir

References

[1] Kaser A, Zeissig S, Blumberg RS. Inflammatory bowel disease. Annual review of immunology. 2009;**28**:573-621

[2] Kaplan GG. The global burden of IBD: From 2015 to 2025. Nature reviews Gastroenterology & Hepatology. 2015;**12**(12):720-727

[3] Nishida A, Inoue R, Inatomi O, Bamba S, Naito Y, Andoh A. Gut microbiota in the pathogenesis of inflammatory bowel disease. Clinical journal of gastroenterology. 2018;**11**(1):1-10

[4] Alshehri D, Saadah O, Mosli M, Edris S, Alhindi R, Bahieldin A. Dysbiosis of gut microbiota in inflammatory bowel disease: Current therapies and potential for microbiota-modulating therapeutic approaches. Bosnian Journal of Basic Medical Sciences. 2021;**21**(3):270

[5] Ng SC, Shi HY, Hamidi N, Underwood FE, Tang W, Benchimol EI, et al. Worldwide incidence and prevalence of inflammatory bowel disease in the 21st century: A systematic review of population-based studies. The Lancet. 2017;**390**(10114):2769-2778

[6] Bouma G, Strober W. The immunological and genetic basis of inflammatory bowel disease. Nature Reviews Immunology. 2003;**3**(7): 521-533

[7] Aldars-García L, Marin AC, Chaparro M, Gisbert JP. The interplay between immune system and microbiota in inflammatory bowel disease: A narrative review. International Journal of Molecular Sciences. 2021;**22**(6):3076

[8] Thoreson R, Cullen JJ. Pathophysiology of inflammatory bowel disease: An overview. Surgical Clinics of North America. 2007;**87**(3):575-585

[9] Seyedian SS, Nokhostin F, Malamir MD. A review of the diagnosis, prevention, and treatment methods of inflammatory bowel disease. Journal of Medicine and Life. 2019;**12**(2):113

[10] Burgmann T, Clara I, Graff L, Walker J, Lix L, Rawsthorne P, et al. The Manitoba inflammatory bowel disease cohort study: Prolonged symptoms before diagnosis—How much is irritable bowel syndrome? Clinical Gastroenterology and Hepatology. 2006;**4**(5):614-620

[11] Yang C, Singh P, Singh H, Le ML, El-Matary W. Systematic review: Thalidomide and thalidomide analogues for treatment of inflammatory bowel disease. Alimentary Pharmacology & Therapeutics. 2015;**41**(11):1079-1093

[12] Cummings JF, Keshav S, Travis SP. Medical management of Crohn's disease. BMJ. 2008;**336**(7652):1062-1066

[13] Tremaine WJ. Diagnosis and treatment of indeterminate colitis. Gastroenterology & Hepatology. 2011;**7**(12):826

[14] Liu JZ, Van Sommeren S, Huang H, Ng SC, Alberts R, Takahashi A, et al. Association analyses identify 38 susceptibility loci for inflammatory bowel disease and highlight shared genetic risk across populations. Nature Genetics. 2015;**47**(9):979-986

[15] De Hertogh G, Aerssens J, Geboes KP, Geboes K. Evidence for the involvement of infectious agents in the pathogenesis of Crohn's disease. World Journal of Gastroenterology: WJG. 2008;**14**(6):845

[16] Kalischuk LD, Buret AG. A role for campylobacter jejuni-induced enteritis in inflammatory bowel disease? American journal of physiology-gastrointestinal and liver. Physiology. 2010;**298**(1):G1-G9

[17] Liu S, Zhao W, Lan P, Mou X. The microbiome in inflammatory bowel diseases: From pathogenesis to therapy. Protein & Cell. 2021;**12**(5):331-345

[18] Hisamatsu T, Kanai T, Mikami Y, Yoneno K, Matsuoka K, Hibi T. Immune aspects of the pathogenesis of inflammatory bowel disease. Pharmacology & Therapeutics. 2013;**137**(3):283-297

[19] Geremia A, Biancheri P, Allan P, Corazza GR, Di Sabatino A. Innate and adaptive immunity in inflammatory bowel disease. Autoimmunity Reviews. 2014;**13**(1):3-10

[20] Bamias G, Cominelli F. Immunopathogenesis of inflammatory bowel disease: Current concepts. Current opinion in gastroenterology. 2007;**23**(4):365-369

[21] Medzhitov R, Preston-Hurlburt P, Janeway CA. A human homologue of the drosophila toll protein signals activation of adaptive immunity. Nature. 1997;**388**(6640):394-397

[22] Poltorak A, Smirnova I, He X, Liu M-Y, Van Huffel C, Birdwell D, et al. Genetic and physical mapping of the LpsLocus: Identification of the Toll-4 receptor as a candidate gene in the critical region. Blood Cells, Molecules, and Diseases. 1998;**24**(3):340-355

[23] Knutson CG, Mangerich A, Zeng Y, Raczynski AR, Liberman RG, Kang P, et al. Chemical and cytokine features of innate immunity characterize serum and tissue profiles in inflammatory bowel disease. Proceedings of the National Academy of Sciences. 2013;**110**(26):E2332-E2E41

[24] Lu Q, Yang M-f, Liang Y-j, Xu J, Xu H-M, Nie Y-Q, et al. Immunology of inflammatory bowel disease: Molecular mechanisms and therapeutics. Journal of Inflammation Research. 2022;**15**:1825

[25] Maloy KJ, Powrie F. Intestinal homeostasis and its breakdown in inflammatory bowel disease. Nature. 2011;**474**(7351):298-306

[26] Medzhitov R, Janeway C Jr. Innate immunity. New England Journal of Medicine. 2000;**343**(5):338-344

[27] Zhang SZ, Zhao XH, Zhang D. Cellular and molecular immuno pathogenesis of ulcerative colitis. Cellular & Molecular Immunology. 2006;**3**(1):35-40

[28] Ghosh S, May MJ, Kopp EB. NF-κB and Rel proteins: Evolutionarily conserved mediators of immune responses. Annual Review of Immunology. 1998;**16**(1):225-260

[29] Caamano J, Hunter CA. NF-κB family of transcription factors: Central regulators of innate and adaptive immune functions. Clinical Microbiology Reviews. 2002;**15**(3):414-429

[30] Blam ME, Stein RB, Lichtenstein GR. Integrating anti–tumor necrosis factor therapy in inflammatory bowel disease: Current and future perspectives. The American Journal of Gastroenterology. 2001;**96**(7):1977-1997

[31] Huang Y, Chen Z. Inflammatory bowel disease related innate immunity and adaptive immunity. American Journal of Translational Research. 2016;**8**(6):2490

[32] Ligumsky M, Simon P, Karmeli F, Rachmilewitz D. Role of interleukin 1 in

inflammatory bowel disease--enhanced production during active disease. Gut. 1990;**31**(6):686-689

[33] Atreya R, Neurath M. New therapeutic strategies for treatment of inflammatory bowel disease. Mucosal Immunology. 2008;**1**(3):175-182

[34] Stockinger B, Veldhoen M. Differentiation and function of Th17 T cells. Current Opinion in Immunology. 2007;**19**(3):281-286

[35] Weaver CT, Harrington LE, Mangan PR, Gavrieli M, Murphy KM. Th17: An effector CD4 T cell lineage with regulatory T cell ties. Immunity. 2006;**24**(6):677-688

[36] Torres M, Rios A. Current view of the immunopathogenesis in inflammatory bowel disease and its implications for therapy. World journal of gastroenterology: WJG. 2008;**14**(13):1972

[37] Leppkes M, Neurath M. Cytokines in inflammatory bowel diseases–update 2020. Pharmacological Research. 2020;**158**:104835

[38] Silva FA, Rodrigues BL, Ayrizono MLS, Leal RF. The immunological basis of inflammatory bowel disease. Gastroenterology Research and Practice. 2016;**2016**:1-11

[39] Kolls JK, Lindén A. Interleukin-17 family members and inflammation. Immunity. 2004;**21**(4):467-476

[40] Harrington LE, Hatton RD, Mangan PR, Turner H, Murphy TL, Murphy KM, et al. Interleukin 17–producing CD4+ effector T cells develop via a lineage distinct from the T helper type 1 and 2 lineages. Nature Immunology. 2005;**6**(11):1123-1132

[41] Ramani K, Biswas PS. Interleukin-17: Friend or foe in organ fibrosis. Cytokine. 2019;**120**:282-288

[42] Aardoom MA, Veereman G, de Ridder L. A review on the use of anti-TNF in children and adolescents with inflammatory bowel disease. International Journal of Molecular Sciences. 2019;**20**(10):2529

[43] Ivanov II, McKenzie BS, Zhou L, Tadokoro CE, Lepelley A, Lafaille JJ, et al. The orphan nuclear receptor RORγt directs the differentiation program of proinflammatory IL-17+ T helper cells. Cell. 2006;**126**(6):1121-1133

[44] Saraiva M, O'garra A. The regulation of IL-10 production by immune cells. Nature Reviews Immunology. 2010;**10**(3):170-181

[45] Duerr RH, Taylor KD, Brant SR, Rioux JD, Silverberg MS, Daly MJ, et al. A genome-wide association study identifies IL23R as an inflammatory bowel disease gene. Science. 2006;**314**(5804):1461-1463

[46] Fuss IJ, Neurath M, Boirivant M, Klein JS, De La Motte C, Strong SA, et al. Disparate CD4+ lamina propria (LP) lymphokine secretion profiles in inflammatory bowel disease. Crohn's disease LP cells manifest increased secretion of IFN-gamma, whereas ulcerative colitis LP cells manifest increased secretion of IL-5. The Journal of Immunology. 1996;**157**(3):1261-1270

[47] Strober W, Fuss IJ. Proinflammatory cytokines in the pathogenesis of inflammatory bowel diseases. Gastroenterology. 2011;**140**(6): 1756-67. e1

[48] MacDonald TT, Biancheri P, Sarra M, Monteleone G. What's the next best cytokine target in IBD? Inflammatory bowel diseases. 2012;**18**(11):2180-2189

[49] Fantini MC, Becker C, Tubbe I, Nikolaev A, Lehr HA, Galle P, et al. Transforming growth factor β induced FoxP3+ regulatory T cells suppress Th1 mediated experimental colitis. Gut. 2006;**55**(5):671-680

[50] Chamouard P, Monneaux F, Richert Z, Voegeli A-C, Lavaux T, Gaub MP, et al. Diminution of circulating CD4+ CD25high T cells in naïve Crohn's disease. Digestive Diseases and Sciences. 2009;**54**(10):2084-2093

[51] Monteleone G, Kumberova A, Croft NM, McKenzie C, Steer HW, MacDonald TT. Blocking Smad7 restores TGF-β1 signaling in chronic inflammatory bowel disease. The Journal of Clinical Investigation. 2001;**108**(4):601-609

[52] Fahlén L, Read S, Gorelik L, Hurst SD, Coffman RL, Flavell RA, et al. T cells that cannot respond to TGF-β escape control by CD4+ CD25+ regulatory T cells. The Journal of Experimental Medicine. 2005;**201**(5):737-746

[53] Miyoshi J, Chang EB. The gut microbiota and inflammatory bowel diseases. Translational Research. 2017;**179**:38-48

[54] Sartor RB. Mechanisms of disease: Pathogenesis of Crohn's disease and ulcerative colitis. Nature Clinical Practice Gastroenterology & Hepatology. 2006;**3**(7):390-407

[55] Ihara S, Hirata Y, Koike K. TGF-β in inflammatory bowel disease: A key regulator of immune cells, epithelium, and the intestinal microbiota. Journal of Gastroenterology. 2017;**52**(7):777-787

[56] Cho JH. The genetics and immunopathogenesis of inflammatory bowel disease. Nature Reviews Immunology. 2008;**8**(6):458-466

[57] Zheng L, Wen X-L. Gut microbiota and inflammatory bowel disease: The current status and perspectives. World Journal of Clinical Cases. 2021;**9**(2):321

[58] Mentella MC, Scaldaferri F, Pizzoferrato M, Gasbarrini A, Miggiano GAD. Nutrition, IBD and gut microbiota: A review. Nutrients. 2020;**12**(4):944

[59] Alameddine J, Godefroy E, Papargyris L, Sarrabayrouse G, Tabiasco J, Bridonneau C, et al. Faecalibacterium prausnitzii skews human DC to prime IL10-producing T cells through TLR2/6/JNK signaling and IL-10, IL-27, CD39, and IDO-1 induction. Frontiers in Immunology. 2019;**10**:143

[60] Patterson AM, Mulder IE, Travis AJ, Lan A, Cerf-Bensussan N, Gaboriau-Routhiau V, et al. Human gut symbiont Roseburia hominis promotes and regulates innate immunity. Frontiers in Immunology. 2017;**8**:1166

[61] Mukherjee A, Lordan C, Ross RP, Cotter PD. Gut microbes from the phylogenetically diverse genus Eubacterium and their various contributions to gut health. Gut Microbes. 2020;**12**(1):1802866

[62] An D, Oh SF, Olszak T, Neves JF, Avci FY, Erturk-Hasdemir D, et al. Sphingolipids from a symbiotic microbe regulate homeostasis of host intestinal natural killer T cells. Cell. 2014;**156**(1-2):123-133

[63] Imaoka A, Shima T, Kato K, Mizuno S, Uehara T, Matsumoto S, et al. Anti-inflammatory activity of probiotic Bifidobacterium: Enhancement of IL-10 production in peripheral blood mononuclear cells from ulcerative

colitis patients and inhibition of IL-8 secretion in HT-29 cells. World Journal of Gastroenterology: WJG. 2008;**14**(16):2511

[64] Bang C, Weidenbach K, Gutsmann T, Heine H, Schmitz RA. The intestinal archaea Methanosphaera stadtmanae and Methanobrevibacter smithii activate human dendritic cells. PLoS One. 2014;**9**(6):e99411

[65] Martinez-Medina M, Garcia-Gil LJ. Escherichia coli in chronic inflammatory bowel diseases: An update on adherent invasive Escherichia coli pathogenicity. World Journal of Gastrointestinal Pathophysiology. 2014;**5**(3):213

[66] Zuo T, Ng SC. The gut microbiota in the pathogenesis and therapeutics of inflammatory bowel disease. Frontiers in Microbiology. 2018;**2247**: 1-13

[67] Shin N-R, Whon TW, Bae J-W. Proteobacteria: Microbial signature of dysbiosis in gut microbiota. Trends in Biotechnology. 2015;**33**(9):496-503

[68] Henke MT, Kenny DJ, Cassilly CD, Vlamakis H, Xavier RJ, Clardy J. Ruminococcus gnavus, a member of the human gut microbiome associated with Crohn's disease, produces an inflammatory polysaccharide. Proceedings of the National Academy of Sciences. 2019;**116**(26):12672-12677

[69] Bashir A, Miskeen AY, Hazari YM, Asrafuzzaman S, Fazili KM. Fusobacterium nucleatum, inflammation, and immunity: The fire within human gut. Tumor Biology. 2016;**37**(3):2805-2810

[70] Engevik M, Danhof H, Britton R, Versalovic J. 20 elucidating the role of FUSOBACTERIUM NUCLEATUM IN intestinal inflammation. Inflammatory Bowel Diseases. 2020;**26**(Supplement_1):S29-S2S

[71] Gringhuis SI, Wevers BA, Kaptein TM, Van Capel TM, Theelen B, Boekhout T, et al. Selective C-Rel activation via Malt1 controls anti-fungal TH-17 immunity by dectin-1 and dectin-2. PLoS Pathogens. 2011;**7**(1):e1001259

[72] Norman JM, Handley SA, Baldridge MT, Droit L, Liu CY, Keller BC, et al. Disease-specific alterations in the enteric virome in inflammatory bowel disease. Cell. 2015;**160**(3):447-460

[73] Lopetuso LR, Ianiro G, Scaldaferri F, Cammarota G, Gasbarrini A. Gut virome and inflammatory bowel disease. Inflammatory Bowel Diseases. 2016;**22**(7):1708-1712

[74] Virgin HW. The virome in mammalian physiology and disease. Cell. 2014;**157**(1):142-150

[75] Frank DN, Amand ALS, Feldman RA, Boedeker EC, Harpaz N, Pace NR. Molecular-phylogenetic characterization of microbial community imbalances in human inflammatory bowel diseases. Proceedings of the National Academy of Sciences. 2007;**104**(34):13780 13785

[76] Kamada N, Seo S-U, Chen GY, Núñez G. Role of the gut microbiota in immunity and inflammatory disease. Nature Reviews Immunology. 2013;**13**(5):321-335

[77] Carvalho FA, Koren O, Goodrich JK, Johansson ME, Nalbantoglu I, Aitken JD, et al. Transient inability to manage proteobacteria promotes chronic gut inflammation in TLR5-deficient mice. Cell host & Microbe. 2012;**12**(2):139-152

[78] Joossens M, Huys G, Cnockaert M, De Preter V, Verbeke K, Rutgeerts P, et al.

Dysbiosis of the faecal microbiota in patients with Crohn's disease and their unaffected relatives. Gut. 2011;**60**(5):631-637

[79] Lloyd-Price J, Arze C, Ananthakrishnan AN, Schirmer M, Avila-Pacheco J, Poon TW, et al. Multi-omics of the gut microbial ecosystem in inflammatory bowel diseases. Nature. 2019;**569**(7758):655-662

[80] Pittayanon R, Lau JT, Leontiadis GI, Tse F, Yuan Y, Surette M, et al. Differences in gut microbiota in patients with vs without inflammatory bowel diseases: A systematic review. Gastroenterology. 2020;**158**(4):930-46. e1

[81] Zhang L, Liu F, Xue J, Lee SA, Liu L, Riordan SM. Bacterial species associated with human inflammatory bowel disease and their pathogenic mechanisms. Frontiers in Microbiology. 2022;**13**

[82] Vakili B, Fateh A, Asadzadeh Aghdaei H, Sotoodehnejadnematalahi F, Siadat SD. Characterization of gut microbiota in hospitalized patients with Clostridioides difficile infection. Current Microbiology. 2020;**77**(8):1673-1680

[83] Shoaei P, Shojaei H, Siadat SD, Moshiri A, Vakili B, Yadegari S, et al. Gut microbiota in burned patients with Clostridioides difficile infection. Burns. 2021

[84] Shoaei P, Shojaei H, Jalali M, Khorvash F, Hosseini SM, Ataei B, et al. Clostridium difficile isolated from faecal samples in patients with ulcerative colitis. BMC Infectious Diseases. 2019;**19**(1):1-7

[85] Shoaei P, Shojaei H, Khorvash F, Hosseini SM, Ataei B, Tavakoli H, et al. Molecular epidemiology of Clostridium difficile infection in Iranian hospitals. Antimicrobial Resistance & Infection Control. 2019;**8**(1):1-7

[86] Azimi T, Nasiri MJ, Chirani AS, Pouriran R, Dabiri H. The role of bacteria in the inflammatory bowel disease development: A narrative review. APMIS. 2018;**126**(4):275-283

[87] Clayton EM, Rea MC, Shanahan F, Quigley EM, Kiely B, Hill C, et al. The vexed relationship between Clostridium difficile and inflammatory bowel disease: An assessment of carriage in an outpatient setting among patients in remission. Official journal of the American College of Gastroenterology| ACG. 2009;**104**(5):1162-1169

[88] Shoaei P, Shojaei H, Khorvash F, Ataei B, Esfandiari Z, Vakili B, et al. Clostridium difficile infection in cancer patients with hospital acquired diarrhea at the teaching hospitals in Iran: Multilocus sequence typing analysis (MLST) and antimicrobial resistance pattern. Annali di Igiene. 2019;**31**(4):365-373

[89] Vakili B, Fateh A, Aghdaei HA, Sotoodehnejadnematalahi F, Siadat SD. Intestinal microbiota in elderly inpatients with Clostridioides difficile infection. Infection and Drug Resistance. 2020;**13**:2723

[90] Monaghan TM, Cockayne A, Mahida YR. Pathogenesis of Clostridium difficile infection and its potential role in inflammatory bowel disease. Inflammatory Bowel Diseases. 2015;**21**(8):1957-1966

[91] Buffie CG, Bucci V, Stein RR, McKenney PT, Ling L, Gobourne A, et al. Precision microbiome reconstitution restores bile acid mediated resistance to Clostridium difficile. Nature. 2015;**517**(7533):205-208

[92] Irving PM, de Lusignan S, Tang D, Nijher M, Barrett K. Risk of common infections in people with inflammatory

bowel disease in primary care: A population-based cohort study. BMJ open Gastroenterology. 2021;**8**(1):e000573

[93] Uchiya K-i, Tomida S, Nakagawa T, Asahi S, Nikai T, Ogawa K. Comparative genome analyses of Mycobacterium avium reveal genomic features of its subspecies and strains that cause progression of pulmonary disease. Scientific Reports. 2017;**7**(1):1-14

[94] Rath T, Roderfeld M, Blöcher S, Rhode A, Basler T, Akineden Ö, et al. Presence of intestinal Mycobacterium avium subspecies paratuberculosis (MAP) DNA is not associated with altered MMP expression in ulcerative colitis. BMC Gastroenterology. 2011;**11**(1):1-19

[95] Quiding-Järbrink M, Smith DA, Bancroft GJ. Production of matrix metalloproteinases in response to mycobacterial infection. Infection and Immunity. 2001;**69**(9):5661-5670

[96] Agrawal G, Hamblin H, Clancy A, Borody T. Anti-mycobacterial antibiotic therapy induces remission in active pædiatric Crohn's disease. Microorganisms. 2020;**8**(8):1112

[97] Feller M, Huwiler K, Schoepfer A, Shang A, Furrer H, Egger M. Long-term antibiotic treatment for Crohn's disease: Systematic review and meta-analysis of placebo-controlled trials. Clinical Infectious Diseases. 2010;**50**(4):473-480

[98] Chen Y, Blaser MJ. Helicobacter pylori colonization is inversely associated with childhood asthma. The Journal of Infectious Diseases. 2008;**198**(4):553-560

[99] Reibman J, Marmor M, Filner J, Fernandez-Beros M-E, Rogers L, Perez-Perez GI, et al. Asthma is inversely associated with helicobacter pylori status in an urban population. PLoS One. 2008;**3**(12):e4060

[100] Kao JY, Rathinavelu S, Eaton KA, Bai L, Zavros Y, Takami M, et al. Helicobacter pylori-secreted factors inhibit dendritic cell IL-12 secretion: A mechanism of ineffective host defense. American Journal of Physiology-Gastrointestinal and Liver Physiology. 2006;**291**(1):G73-G81

[101] Rad R, Brenner L, Bauer S, Schwendy S, Layland L, da Costa CP, et al. CD25+/Foxp3+ T cells regulate gastric inflammation and helicobacter pylori colonization in vivo. Gastroenterology. 2006;**131**(2):525-537

[102] Jafari F, Shokrzadeh L, Dabiri H, Baghaei K, Yamaoka Y, Zojaji H, et al. vacA genotypes of helicobacter pylori in relation to cagA status and clinical outcomes in Iranian populations. Japanese Journal of Infectious Diseases. 2008;**61**(4):290

[103] Bohr UR, Glasbrenner B, Primus A, Zagoura A, Wex T, Malfertheiner P. Identification of enterohepatic helicobacter species in patients suffering from inflammatory bowel disease. Journal of Clinical Microbiology. 2004;**42**(6):2766-2768

[104] Basset C, Holton J, Bazeos A, Vaira D, Bloom S. Are helicobacter species and enterotoxigenic Bacteroides fragilis involved in inflammatory bowel disease? Digestive Diseases and Sciences. 2004;**49**(9):1425-1432

[105] Zhang L. Oral campylobacter species: Initiators of a subgroup of inflammatory bowel disease? World Journal of Gastroenterology: WJG. 2015;**21**(31):9239

[106] Man SM, Zhang L, Day AS, Leach ST, Lemberg DA, Mitchell H.

Campylobacter concisus and other campylobacter species in children with newly diagnosed Crohn's disease. Inflammatory Bowel Diseases. 2010;**16**(6):1008-1016

[107] Zhang L, Lee H, Grimm MC, Riordan SM, Day AS, Lemberg DA. Campylobacter concisus and inflammatory bowel disease. World Journal of Gastroenterology: WJG. 2014;**20**(5):1259

[108] Nielsen HL, Nielsen H, Ejlertsen T, Engberg J, Günzel D, Zeitz M, et al. Oral and fecal campylobacter concisus strains perturb barrier function by apoptosis induction in HT-29/B6 intestinal epithelial cells. PLoS One. 2011;**6**(8):e23858

[109] Liu H, Hong XL, Sun TT, Huang XW, Wang JL, Xiong H. Fusobacterium nucleatum exacerbates colitis by damaging epithelial barriers and inducing aberrant inflammation. Journal of Digestive Diseases. 2020;**21**(7):385-398

[110] Darfeuille-Michaud A, Boudeau J, Bulois P, Neut C, Glasser A-L, Barnich N, et al. High prevalence of adherent-invasive Escherichia coli associated with ileal mucosa in Crohn's disease. Gastroenterology. 2004;**127**(2):412-421

[111] Dreux N, Denizot J, Martinez-Medina M, Mellmann A, Billig M, Kisiela D, et al. Point mutations in FimH adhesin of Crohn's disease-associated adherent-invasive Escherichia coli enhance intestinal inflammatory response. PLoS Pathogens. 2013;**9**(1):e1003141

[112] Lapaquette P, Bringer MA, Darfeuille-Michaud A. Defects in autophagy favour adherent-invasive Escherichia coli persistence within macrophages leading to increased pro-inflammatory response. Cellular Microbiology. 2012;**14**(6):791-807

[113] Hill C, Guarner F, Reid G, Gibson GR, Merenstein DJ, Pot B, et al. Expert consensus document: The international scientific Association for Probiotics and Prebiotics consensus statement on the scope and appropriate use of the term probiotic. Nature reviews Gastroenterology & Hepatology. 2014;**11**:506-514

[114] Abraham BP, Quigley EM. Probiotics in inflammatory bowel disease. Gastroenterology Clinics. 2017;**46**(4):769-782

[115] Mishra J, Stubbs M, Kuang L, Vara N, Kumar P, Kumar N. Inflammatory bowel disease therapeutics: A focus on probiotic engineering. Mediators of Inflammation. 2022:**2022**

[116] Akram W, Garud N, Joshi R. Role of inulin as prebiotics on inflammatory bowel disease. Drug discoveries & therapeutics. 2019;**13**(1):1-8

[117] Colombel J-F, Shin A, Gibson PR. AGA clinical practice update on functional gastrointestinal symptoms in patients with inflammatory bowel disease: Expert review. Clinical Gastroenterology and Hepatology. 2019;**17**(3):380-90. e1

[118] Jakubczyk D, Leszczyńska K, Górska S. The effectiveness of probiotics in the treatment of inflammatory bowel disease (IBD)—A critical review. Nutrients. 2020;**12**(7):1973

[119] Doron S, Snydman DR. Risk and safety of probiotics. Clinical Infectious Diseases. 2015;**60**(suppl_2):S129-SS34

[120] FreedmanSB,Williamson-UrquhartS, Farion KJ, Gouin S, Willan AR, Poonai N, et al. Multicenter trial of a

combination probiotic for children with gastroenteritis. New England Journal of Medicine. 2018;**379**(21):2015-2026

[121] Bjarnason I, Sission G, Hayee BH. A randomised, double-blind, placebo-controlled trial of a multi-strain probiotic in patients with asymptomatic ulcerative colitis and Crohn's disease. Inflammopharmacology. 2019;**27**(3):465-473

[122] Bian X, Wu W, Yang L, Lv L, Wang Q, Li Y, et al. Administration of Akkermansia muciniphila ameliorates dextran sulfate sodium-induced ulcerative colitis in mice. Frontiers in Microbiology. 2019;**2259**:1-18

[123] Wang L, Tang L, Feng Y, Zhao S, Han M, Zhang C, et al. A purified membrane protein from Akkermansia muciniphila or the pasteurised bacterium blunts colitis associated tumourigenesis by modulation of CD8+ T cells in mice. Gut. 2020;**69**(11):1988-1997

[124] Eindor-Abarbanel A, Healey GR, Jacobson K. Therapeutic advances in gut microbiome modulation in patients with inflammatory bowel disease from pediatrics to adulthood. International Journal of Molecular Sciences. 2021;**22**(22):12506

[125] Lee J-Y, Cevallos SA, Byndloss MX, Tiffany CR, Olsan EE, Butler BP, et al. High-fat diet and antibiotics cooperatively impair mitochondrial bioenergetics to trigger dysbiosis that exacerbates pre-inflammatory bowel disease. Cell host & Microbe. 2020;**28**(2):273-84. e6

[126] Oka A, Sartor RB. Microbial-based and microbial-targeted therapies for inflammatory bowel diseases. Digestive Diseases and Sciences. 2020;**65**(3):757-788

[127] Roda G, Jharap B, Neeraj N, Colombel J-F. Loss of response to anti-TNFs: Definition, epidemiology, and management. Clinical and Translational Gastroenterology. 2016;**7**(1):e135

[128] Paramsothy S, Paramsothy R, Rubin DT, Kamm MA, Kaakoush NO, Mitchell HM, et al. Faecal microbiota transplantation for inflammatory bowel disease: A systematic review and meta-analysis. Journal of Crohn's and Colitis. 2017;**11**(10):1180-1199

[129] Hvas CL, Jørgensen SMD, Jørgensen SP, Storgaard M, Lemming L, Hansen MM, et al. Fecal microbiota transplantation is superior to fidaxomicin for treatment of recurrent Clostridium difficile infection. Gastroenterology. 2019;**156**(5):1324-32. e3

[130] Khoruts A, Sadowsky MJ. Understanding the mechanisms of faecal microbiota transplantation. Nature reviews Gastroenterology & Hepatology. 2016;**13**(9):508-516

[131] Cammarota G, Ianiro G, Kelly CR, Mullish BH, Allegretti JR, Kassam Z, et al. International consensus conference on stool banking for faecal microbiota transplantation in clinical practice. Gut. 2019;**68**(12):2111-2121

[132] Draper LA, Ryan FJ, Smith MK, Jalanka J, Mattila E, Arkkila P, et al. Long-term colonisation with donor bacteriophages following successful faecal microbial transplantation. Microbiome. 2018;**6**(1):1-9

Chapter 3

Gut Microbiota Potential in Type 2 Diabetes

Shahzad Irfan, Humaira Muzaffar, Haseeb Anwar and Farhat Jabeen

Abstract

Appropriate metabolic regulation is vital for health. Multiple factors play important roles in maintaining the metabolic system in different physiological conditions. These factors range from intestinal metabolism of food and absorption of nutrients, pancreatic hormones and their interplay under feeding and fasting, hepatic regulation of macronutrient formation and metabolism storage of macronutrients in skeletal muscles. Intestinal metabolism of ingested food and subsequent nutrient absorption depends on the symbiotic microbial community residing in the gut. The specific ratio of different microbial phyla in the gut has proved to be extremely important for the beneficial role of the gut microbiome. The importance of gut microbiome in the regulation of metabolism has been highlighted with reports of the abnormal ratio of gut microbial community resulting in different metabolic disturbances ranging from obesity to the development of diabetes mellitus. The physiological impact of insulin on the metabolic regulation of macronutrients has recently been shown to be augmented by the secondary metabolites produced by anaerobic fermentation. The current chapter aims to highlight recent findings in the regulation of extraintestinal metabolism by gut microbiome with a specific emphasis on the physiology and pathophysiology of the pancreas in health and disease.

Keywords: Gut microbiota, diabetes mellitus, probiotics, pancreas

1. Introduction

Insulin is predominantly the most important endogenous protein responsible for the physiological regulation of metabolism [1]. Exogenous insulin is the only substantial treatment option for patients suffering from insulin deficiency since the initial discovery of insulin by Sir Frederick G Banting and its purification by James B. Collip in 1921 [2, 3]. The pancreatic gland is responsible for the regulated secretion of insulin to maintain glucose homeostasis under different physiological conditions [4, 5]. Islets of Langerhans present in the pancreas contain cells that secret specific hormones which help in maintaining glucose levels during feeding and fasting [5–9]. Islets of Langerhans are defined as closed areas containing multiple cell types with enormous vascular and nervous innervation [5]. Islets of Langerhans are designated as the endocrine portion of the pancreas. The exocrine part of the pancreas surrounds islets of Langerhans. Different cell types present in the islets secrete different types of hormones. Islets contain four different types of endocrine cells: alpha (α) cells (glucagon), beta (β) cells (insulin), delta (δ) cells (somatostatin) and PP cells (pancreatic polypeptide) [5]. Alpha cells are responsible for the secretion of glucagon hormone to enhance blood glucose levels under fasting conditions while β cells are responsible for insulin secretion which initiates postprandial glucose metabolism and thus controls the rising blood glucose levels [10, 11]. Apart from glucose metabolism, insulin is also involved in lipid and protein metabolism [12–14]. Blood glucose level acts as the main trigger for the release of insulin from β cells, a phenomenon known as glucose-stimulated insulin secretion (GSIS) [15–17]. Glucose at normal physiological levels not only induces insulin gene transcription by recruiting transcription factors (PDX-1, MafA and NeuroD) but also improves the insulin mRNA stability thus acting as a major physiologic regulator of insulin [18–20]. Glucose enters the β cells via glucose-specific channels present on the cell membrane commonly known as glucose transporters (GLUT) [4, 17]. Numerous types of glucose transporters are present in different tissue of the body [21]. But specifically, GLUT2 is most abundant and functional in the pancreas (β cells) and liver (hepatocytes) whereas GLUT4 is present on skeletal and cardiac muscles and adipocytes [21, 22].

2. Insulin and macronutrient metabolism

Insulin is the primary hormone responsible for initiating carbohydrate metabolism through phosphorylation of glucose and subsequent formation of glucose-6-phosphate inside the cells [23, 24]. Insulin activates the hexokinase enzymes in non-hepatic tissues and glucokinase (GCK) in β cells and hepatocytes to initiate glucose phosphorylation [25–28]. The insulin hormone acts by binding to the cell surface insulin receptors which are vastly distributed in different tissues of the body [29]. The binding of insulin to its receptors activates adaptor proteins known as insulin receptor substrates (e.g. IRS1, IRS2) [30]. IRS protein converts the tyrosine phosphorylation signal into the lipid kinase by activating phosphoinositide2-kinase enzyme (PI3K). Activated PI3K further recruits ATP molecules which activates AKT (serine and threonine kinase) [31]. The Discovery of insulin's primary role in activating AKT proved a landmark in explaining the conversion of tyrosine phosphorylation into serine/threonine phosphorylation signal. AKT activation also explains the insulin induced regulation of key steps in insulin signaling including (a) glucose uptake by glucose transporter (GLUT4), (b) glycogen synthesis by glycogen synthase kinase 3 (GSK3) inhibition, (c) synthesis of protein and fats via activation of the mechanistic target of rapamycin (mTOR), (d) gene expression regulation at the transcriptional levels by forkhead family box O (FOXO) transcription factor proteins. Insulin enhances GLUT4 activity in muscle and adipose tissue thus increasing the rate of glucose transport, glycolysis and subsequent glycogen synthesis in these tissues [32]. Insulin also prevents hepatic glucose synthesis by inhibiting hepatic glycogenolysis and gluconeogenesis [33–35].

Apart from glucose metabolism, insulin influences lipid and protein metabolism through multiple means. Insulin lowers the plasma fatty acid levels by decreasing adipocyte lipolysis and enhancing the hepatic formation of very low density lipoprotein (VLDL) [36–41]. Insulin increases the protein synthesis in skeletal muscles and the liver by enhancing the amino acid transport inside the cells and reducing protein degradation and urea formation [13, 42–47]. These metabolic effects of insulin on carbohydrates, lipids and proteins highlight the importance of insulin signaling in maintaining a nutritional consistency at the cellular level and ensuring a balanced physiological interplay between multiple tissues under diverse physiological conditions.

3. Glucose homeostasis: insulin-glucagon interplay

Alpha and β cells work together to maintain glucose homeostasis under feeding and fasting conditions through the periodic release of insulin and glucagon respectively [4, 25, 48]. Feeding results in increased plasma glucose levels. Rising plasma glucose levels demand immediate systemic activation of glucose metabolism by insulin. A delayed or deficient activation of glucose metabolism will result in abnormally high plasma and cellular glucose levels, a medical condition known as hyperglycemia. Glucose at higher-than-normal concentrations induces glucotoxic effects inside the cells. Rising postprandial glucose levels will trigger β cells to synthesize and secrete insulin. The postprandial rise in insulin levels activates glucokinase and hexokinase activity resulting in glucose phosphorylation in hepatocytes and muscle cells. Conversion of glucose into glucose-6-phosphate will result in the decline of plasma glucose levels over time. Physiologically because of GSIS the postprandial rise in the insulin secretion from β cells declines over time as the blood glucose level decline [15]. Thus, the rising plasma glucose levels provide positive feedback to enhance insulin secretion and the declining plasma glucose levels act as a negative feedback loop to lower insulin levels. Insulin negatively impacts glucagon secretion [49–52]. Plasma glucose levels decline under fasting conditions. As glucose is the primary cellular source to generate ATP, a minimum threshold of plasma glucose levels must be maintained to avoid hypoglycemia.

Hypoglycemia is a serious medical condition characterized by very low plasma glucose levels. Fasting induced a decline in plasma glucose levels and subsequent diminished insulin levels initiate glucagon synthesis and secretion from alpha cells. To avoid hypoglycemia during fasting, glucagon enhances plasma glucose levels by activating hepatic gluconeogenesis/glucogenolysis thus forming glucose molecules from non-carbohydrate sources [10, 53–57]. Glucagon secretion from alpha cells and insulin secretion from β cells are also regulated by incretin hormones secreted from the intestines [58]. Incretin hormones are gut peptides secreted from the L and K cells of the small intestine and include glucose-dependent insulinotropic polypeptide (GIP) and glucagon-like peptide-1 (GLP-1) [59, 60]. GIP and GLP-1 functional receptors are present on both alpha and β cells. In normal physiological conditions, the GIP induces glucagon secretion from alpha cells during fasting or hypoglycemic state whereas GLP-1 induces insulin secretion from β cells and inhibits glucagon secretion from alpha cells [59, 61].

4. Diabetes mellitus

Diabetes mellitus is primarily a metabolic dysfunction resulting in a significant reduction in the cellular ability to metabolize glucose because of either the lack of insulin or insulin inactivity (insulin resistance) [62, 63]. Diabetes mellitus is expected to affect 700 million worldwide by 2040 [64]. The compromised ability of the cells to metabolize glucose results in increased cellular and plasma levels of glucose, a condition known as hyperglycemia. Hyperglycemia induces tissue damage mainly through the increased influx of glucose through the polyol pathway and increased formation of advanced glycation end products (AGEs) and subsequent increased expression of AGE receptors and their ligand [65, 66]. Overproduction of reactive oxygen species (ROS) due to hyperglycemia through mitochondria acts as the main trigger for the activation of the polyol pathway, formation of AGEs and increase in AGE receptor expression [67].

Diabetes mellitus has been categorized in two primary forms: Type 1 Diabetes Mellitus (T1DM) and Type 2 Diabetes Mellitus (T2DM). T1DM has been characterized by a mutation in the insulin gene or immune cell-mediated destruction of β cell resulting in either the synthesis of abnormal insulin protein that fails to activate insulin receptors or a complete lack of endogenous insulin secretion [63]. T1DM patients are usually diagnosed early in their life. The only possible medical treatment referred to these patients is the multiple daily doses of synthetic insulin. T2DM on the other hand is much more complicated and requires a thorough diagnostic approach [62, 68–70]. T2DM is considered one of the most common metabolic disorders globally. Major risk factors for T2DM include a sedentary lifestyle, lack of exercise, excessive use of a high-carb and high-fat diet, overweight and obesity [71]. Poor lifestyle and dietary habits have been attributed to the global incidence of type 2 diabetes in the last 2 decades. Obesity, visceral fat deposition and increased body mass index (BMI) play a central role in the pathophysiology of type 2 diabetic patients [62]. Quality, quantity and type of food have been debated to be the primary cause of this global incident. A healthy diet with the appropriate amount of nutrients and fiber and a certain level of physical activity has been advised globally to counter the incidence of T2DM in young adults.

The development of T2DM is mainly caused by the significant decline in insulin secretion from β cells or the inability of insulin-responsive tissues (muscles, fat and liver) to respond to insulin, mainly because of defective insulin signaling resulting in hyperinsulinemia and subsequent insulin resistance [72–75]. Failure of the insulin hormone to activate insulin receptors at the cellular level has been attributed to be the major cause of hyperinsulinemia and insulin resistance [76, 77]. Insulin binding to insulin receptors at the plasma membrane activates a signaling cascade that initiates glucose metabolism inside the cells. Insulin-bound insulin receptors or activated insulin receptors go through internalization at the plasma membrane, a phenomenon known as insulin receptor endocytosis [1, 78]. Following the activation, the endocytosis of the insulin receptor is the primary physiological mechanism through which the duration and intensity of insulin signaling are controlled. Hyperinsulinemia accelerates insulin receptor endocytosis and affects the presence of adequate functional insulin receptors at the plasma membrane resulting in insulin resistance [79]. Apart from accelerated insulin receptor endocytosis, insulin-stimulated insulin receptor kinase activity is also decreased in diabetic patients [80]. Compromised insulin signaling fails to activate glucose metabolic enzymes like glucokinase and hexokinase resulting in hyperglycemia. High plasma glucose levels initiate glucose-stimulated insulin secretory (GSIS) response from β cells resulting in the rise of plasma insulin levels. The rising insulin levels should be normalized over time because of the renal insulin clearance mechanism. But compromised renal insulin clearance rate in diabetic subjects results in abnormally high plasma levels of insulin (hyperinsulinemia) [81, 82].

Hyperinsulinemia and hyperglycemia in theory cannot trigger alpha cells to secrete glucagon. But it has been observed that T2DM patients with insulin resistance, hyperinsulinemia and hyperglycemia also have abnormally high plasma levels of glucagon [83]. Hinting toward the disturbance in the alpha and β cell interplay through the inability of the insulin to block glucagon gene transcription [84]. T2DM is also characterized by a decrease in GLP-1 secretion from L cells of the small intestine [85, 86]. Indicating a pathophysiological role of the gut in the development and progression of type 2 diabetes [87, 88]. GLP-1 receptor agonists which induce an increase in insulin secretion from β cells and inhibit glucagon secretion are the major treatment option for T2DM patients to combat hyperglycemic conditions [89–91].

5. Gut microbiota profile in Type 2 diabetes mellitus

The gut microbiome was first defined scientifically in 2001 as "an ecological community of commensal, symbiotic and pathogenic microorganisms that collectively share our body space" [92]. Approximately 100 trillion microbes are found in the human gastrointestinal tract (GIT) and strongly influence the health status of individuals either directly or indirectly [93–97]. The primary reason for the pathophysiological effect of the gut microbiome on human physiology has been attributed to the disruption of the stable communities of gut microbes through medication, diet and lifestyle. A normal, healthy gut microbiome profile is termed eubiosis and abnormal gut microbiome composition is called dysbiosis [98–106]. Eubiosis typically refers to an ideal bacterial population comprising 95% of Bacteroidetes and 5% Firmicutes producing abundant microbial metabolites like short-chain fatty acids (SCFAs), branched-chain amino acids (BCAAs) and impacting lipid metabolism. SCFAs like butyrate, acetate and propionate are produced by the anaerobic fermentation of non-digestible carbohydrates (dietary fiber) and promote gut integrity and protect gut epithelial lining by forming tight junctions and preventing gut permeability [107]. These microbial secondary metabolites act as central components in microbe to host signaling pathways activation. Much of the specific microbiota involved in the production of these important secondary metabolites are reduced in T2DM patients.

Substantial data from human studies support the possibility that dysbiosis triggers obesity, inflammation, insulin resistance and T2DM [108–111]. Association of dysbiosis is also attributed to the pathogenesis of intestinal tissue. Intestinal disorders attributed to dysbiosis include inflammatory bowel disease, irritable bowel syndrome (IBD) and coeliac disease [102, 111, 112]. Whereas metabolic syndrome, obesity, and cardiovascular complications are attributed as extra-intestinal effects of dysbiosis. Dysbiosis has also been attributed not only to the initiation of the T2DM in humans (a condition known as prediabetes) but also during the progression and subsequent secondary complications of T2DM with several lines of evidence suggesting that manipulation of the gut microbiome helps to minimize or alleviate the T2DM conditions [98, 113–121].

The role of gut microbiota in health and disease and specifically the pathogenesis of T2DM has been experimentally investigated mainly by using rodent models as a limited amount of experimental data can be generated through human studies. Keeping in mind that the rodents and human physiology are not exactly similar and certain physiological differences exist. The non-human primates seem to be a much more appropriate animal model to study different aspects of primate physiology including the gut microbiome and its interaction with metabolic dysregulation [122–127]. Nonetheless, the current understanding of the role of the gut microbiome in the context of metabolic syndrome or pathogenesis of diabetes mellitus has primarily originated from the data on rodent and human studies [94, 97, 116, 128–133]. Interestingly efforts have been made in the past to characterize the gut microbiome in normal and diabetic individuals as well as some therapeutic approaches have been adopted [95, 98, 113, 116, 117, 120, 121, 134].

The attempts to characterize the normal human gut microbiome revealed four primary phyla which are responsible for the physiological role of gut in metabolic modulation [128, 132, 133, 135–141]. These four specific phyla/families of microbes present in the gut include Bacteroidetes (Bacteroidota), Firmicutes (Bacillota), Proteobacteria (Pseudomonadota) and actinobacteria (Actinomycetota) [95, 142]. The specific proportion for each of these phyla in normal physiological and homeostatic conditions indicates that the largest group of microbes is the Firmicutes which make up to 64% of the total gut microbiota. Followed by the Bacteroidetes, which make up the second-largest group, contributing up to 23% of the total gut microbiota. Proteobacteria and actinobacteria contribute the rest with 8% and 3% respectively. These specific percentage contributions of each phylum are extremely important physiologically. Increased prevalence of pro-inflammatory conditions such as obesity, T2DM, arthritis and even cancer have been attributed to the disruption of these specific percentage contributions of each phylum [132, 143]. Human and animal data have highlighted the unique compositional changes in the microbiota profiles at the phylum level in T2DM conditions [113, 128]. T2DM patients exhibit increased membrane transport of sugars, BCAA transportation, methane metabolism and sulfate reduction [128]. These patients also have reduced butyrate biosynthesis and cofactors/vitamins metabolism.

Although a certain level of discrepancy does exist in terms of phyla composition data between different T2DM patients which has been attributed to the specific geographical location, culture-specific diet and medication use [144]. Numerous independent research groups have reported widely contrasting microbiota findings in the context of phyla composition in T2DM patients [113, 114, 117, 119, 128, 145, 146]. It seems highly unlikely that a single microbe species can play a significant or dominant role in determining the risk of T2DM. The conflicting data from several independent groups also have some interesting similarities. Specifically, it was a common observation among T2DM patients that butyrate-producing microbes were particularly depleted [117, 128]. As human microbiome is comprised mainly of Bacteroidetes and Firmicutes with a specific ratio (B/F > 1) and obesity has been shown to impact this ratio and result in the increased prevalence of Firmicutes to that of Bacteroidetes [109, 147–149]. Implicating that a disrupted B/F ratio can contribute to obesity in humans. Similarly increased concentration of Bacteroidetes and Proteobacteria with a significant decline in Firmicutes has been reported in T2DM patients [113] T2DM also demonstrates an increase in pathogenic microbial species like *Clostridium symbiosum*, *Clostridium ramosum*, and *Escherichia coli* resulting in systemic inflammation [119, 128].

Insulin resistance has also been attributed to disrupted Bacteroidetes and Firmicutes (B/F) ratio. An altered B/F ratio impacts intestinal permeability and lipopolysaccharide (LPS) from proteobacteria are translocated from inside the gut. LPS translocation activates immune response through interleukin-1 (IL-1), tumor necrosis factor (TNF), Jun N-terminal kinases (JNK) and IkB kinase (IKK). LPS-induced activation of JNK and IKK results in the phosphorylation of insulin receptor substrate (IRS) which fails to activate downstream effector molecules like PI3K and AKT thus rendering the insulin signaling cascade ineffective [150, 151]. IKK also activates the nuclear translocation of nuclear factor kappa B (NF-kB). NF-kB, a transcription factor, induces the expression of several genes involved in inflammatory and apoptotic responses [152–155]. The inflammatory state also called metabolic endotoxemia is accompanied by insulin resistance and obesity.

6. Regulation of glucose homeostasis by Gut microbiota

Gut microbiota has been shown to impact the pancreas directly. Gut microbiota has been proposed to modulate glucose homeostasis through multiple mechanisms [115, 116, 118, 156, 157]. Experimental data support four specific mechanisms through which the gut microbiome influences glucose homeostasis; (1) the β cell modulating effects of metabolites that are formed due to gut anaerobic microbial fermentation [157–159], (2) induction of cytokine activity in the islets of Langerhans via inflammatory cascades [160–164], (3) direct islets signaling affecting insulin and glucagon secretion through incretins modulation [87, 165], (4) alteration in the gut permeability, thus permitting the influx of toxins through intestinal mucosal barrier [166]. Mechanisms 1 and 3 are mainly considered for increased T2DM susceptibility and whereas mechanisms 2 and 4 are particularly implicated in the development of T1DM in early life. As T1DM is characterized by a significant reduction in the number of functional β cells. Cytokine and toxin-induced β cell apoptosis or dedifferentiation are considered major risk factors for T1DM.

Abnormal gut microbiome composition alters the intestinal barrier which favors absorption and increased circulating levels of LPS and BCAA. LPS induces low-grade inflammation and insulin resistance while BCAA is associated with an increased risk of T2DM development. An altered intestinal barrier also reduces the absorption of beneficial SCFAs and secondary bile acids. Metabolically SCFAs are mainly produced as an energy source for the gut epithelium. Butyrate is used by colonic epithelial cells for energy, acetate is used as a fatty acid precursor like cholesterol and propionate is a precursor for the process of hepatic gluconeogenesis [167–169]. Animal data have shown the beneficial impact of acetate supplementation on insulin resistance and glucose tolerance in animals fed with a high-fat diet [170]. Acetate at high intravenous (*i.v*) dose has also been reported to acutely enhance circulating levels of GLP-1 in humans [171]. Butyrate supplementation has been reported to enhance insulin sensitivity in mice fed with a high-fat diet while obesity and insulin resistance fail to develop over the course of 16 weeks [165].

Functional modulation of β cells through secondary metabolites is highly important in maintaining homeostatic glucose levels. SCFA has been highlighted as an important signaling molecule as the recent findings of the presence of functional SCFA cell surface receptors on different tissues including gut and peripheral tissues [172–175]. Gut microbial modulation of the host's metabolism modulated by SCFA production has been demonstrated by the activation of G-protein coupled cell surface receptors (*GPCRs*) also known as free fatty acid receptors (*FFAR*). FFAR includes different GPCRs which bind fatty acids of different chain lengths. GPR40 (FFA1), GPR84, and GPR 120 (FFA4) bind with the medium and long-chain fatty acids. Whereas GPR43 (FFA20), GPR41 (FFA3), and GPR109 bind with SCFAs. Propionate and acetate are found to be the most potent agonists of FFA3/GPR41 while butyrate selectively binds with FFA2. FFARs have been shown to be present in different peripheral tissues including the gut, liver, and pancreas. SCFAs like butyrate and propionate along with secondary bile acids indirectly modulate β cells. SCFAs enhance insulin secretion by activating GLP-1 secretion from the intestines [158, 176]. Butyrate and propionate bind and activate G-protein coupled receptors (GPR43, GPR119) present on the enteroendocrine L cells and stimulate the release of GLP-1 in humans [177, 178]. Propionate also has been reported to influence β cell activity directly in humans. Propionate inhibits inflammatory cytokine-induced β cell apoptosis in human islets and enhances GSIS response from β cells independent of increased GLP-1 levels [179]. FFARs have been expressed by β cells and reported to modulate β cell activity in terms of GSIS [174, 180]. Apart from β cell modulation in terms of GSIS response, an interesting observation was made in these studies that high-fat diet-induced insulin resistance in a mouse model has shown to influence FFA2 receptor expression in β cells. Apart from these above-mentioned *in vivo* studies a recently published *in vitro* data further extends the notion that acetate, propionate, and butyrate separately enhance insulin secretion along with an increase in the expression levels of insulin genes from rat islets during long-term incubation [181]. Interestingly the authors noted that long-term incubation with butyrate induced a significant downregulation of β cell-specific key transcriptional factors and functional genes involved in the maturity-onset diabetes of the young (MODY). Another interesting finding which was made in this recent study was the significant suppression of the β cell identity genes like *GLUT2*, *GCK*, *Pdx1*, *MafA*, *Nkx-6.1*, and *NeuroD1* after the long-term incubation with butyrate. The global suppression of the β cell identity gene was surprisingly independent of the deacetylase activity of butyrate indicating a non-DNA acetylation mechanism involved. A significant decrease in the gene expression pattern of GLUT2 and GCK in rat islets after the long-term incubation with butyrate indicates that glucose or GSIS was not involved in the increased mRNA levels of *INS1* and secretion of insulin protein. Instead the basal levels of intracellular calcium ions $[Ca^{2+}]_i$ was much higher in butyrate-treated islets as compared to the control. The combined effect of acetate, propionate, and butyrate on isolated rat or mouse islets in short- and long-term incubations needs to be examined in future studies. Along with *in vivo* approaches to fully characterize the impact of microbial metabolites on glucose homeostasis in different physiological conditions.

7. Prebiotics, probiotics and diet

Prebiotics are food ingredients that are non-digestible but fermentable oligosaccharides. The primary role of prebiotics in food is to stimulate the fermenting activity of gut microbes and eventually trigger the growth of beneficial gut microbes [182–185]. Probiotics on the other hand are special foods that contain a certain amount of alive non-pathogenic bacteria which help to improve gut health and confer eubiosis [186, 187]. Bifidobacteria, lactobacilli streptococci and *E. coli* are the main bacterial strains that constitute most of the available probiotics. Prebiotics and probiotics supplementation has been shown to reduce inflammation and obesity in T2D patients [188–192]. The beneficial effect of prebiotics and probiotics on gut health, in general, is well accepted. However, the expected benefits of pre and probiotics in dysbiotic T2D patients is limited. Advanced stage or elderly diabetic patients fails to respond to pre and probiotic supplementation as compared to young and early-stage T2D patients [190, 193, 194]. An appropriate ratio of different gut microbes is extremely important for proper metabolic physiology. As dysbiosis has been attributed as an important permissive or causative factor in developing T2D. Extensive use of high-fat diets, diets which are also called western diets or fast food, have the ability to modulate gut microbiota. Specifically, the downregulation of beneficial Bifidobacteria

(Actinobacteria), which helps to break down food, and nutrient absorption and helps to alleviate constipation and diarrhea by fighting off pathogenic microbes. As Actinobacteria only makes up to 3% of the total gut microbiota, the high fat diet-induced decrease in Bifidobacteria causes an acute pathological impact on metabolism and gut health. A high-fat diet also induces an unwanted increase in the proteobacteria, which usually accounts for a maximum 8%. An increase in the LPS containing proteobacteria causes inflammation and obesity, a condition known as endotoxemia which is accompanied by insulin resistance. Oligofructose-containing prebiotics has been shown to lower LPS containing Proteobacteria by enhancing Bifidobacterial thus modulating endotoxemia and via GLP-1 dependent pathway improves glucose tolerance [195]. Prebiotics, probiotics and fecal microbial transplantation (FMTs) are the main treatment options to enhance gut microbiota. The limited success of these treatment options to restore and maintain the eubiosis over time and the unique microbiology of the gut microbiome has called for a better understanding of gut microbial response and adaptation to different diets and lifestyles. An elegant recent report documenting the *in vivo* bacterial gene expression profile in the gut in different groups of mice indicates that bacterial gene expression is hugely impacted by the type of food present in the gut [196]. Lifestyles, cultures and specific diets have been shown to modulate the gut microbiome in healthy non-diabetic subjects (**Figure 1**). These are lifestyle/diet-induced effects that eventually cause metabolic disorders and obesity. Apart from these detrimental outcomes of certain lifestyles, medications especially antibiotics can severely damage the overall population and the specific ratio of the gut microbiome.

Figure 1.
Impact of the use of probiotics and prebiotics on the gut microbiome in terms of its functionality and improving the glycemic control through manipulation of multiple factors like improved incretin secretions, increase in the production of SCFAs, improved bile acid metabolism and the decrease in the LPs induced low grade inflammatory response.

8. Conclusion

Diabetes mellitus has emerged as the major metabolic disorder in the last two decades. A sedentary urban lifestyle, increased consumption of processed and fried foods and diets high in fat and protein have been indicated as the main reason for unhealthy weight gain causing obesity and disrupting the normal physiological pathways responsible for metabolic homeostasis. The role of the gut microbiome in ensuring a healthy metabolic and immune system is paramount. The remarkable research efforts made in the last two decades highlight gut microbial imbalance or dysbiosis as a common finding in diabetic patients. The direct and indirect regulatory influence of the gut microbial activity on the islet's functionality has been experimentally characterized in rodent models. The experimental findings highlight the importance of a healthy gut microbial community and the use of the appropriate amount of dietary fiber to support fermentation and production of beneficial SCFAs which not only impact the intestinal permeability but also influence β cell activity directly as well as indirectly. The use of pre or probiotics along with a healthy diet comprising enough dietary fiber is a prerequisite for communities and individuals suffering from obesity and diabetes.

Author details

Shahzad Irfan[1*], Humaira Muzaffar[1], Haseeb Anwar[1] and Farhat Jabeen[2]

1 Department of Physiology, Government College University, Faisalabad, Pakistan

2 Department of Zoology, Government College University, Faisalabad, Pakistan

*Address all correspondence to: shahzadirfan@gcuf.edu.pk

References

[1] Petersen MC, Shulman GI. Mechanisms of insulin action and insulin resistance. Physiological Reviews. 2018;**98**(4):2133-2223

[2] Hegele RA, Maltman GM. Insulin's centenary: The birth of an idea. The Lancet Diabetes and Endocrinology. 2020;**8**(12):971-977

[3] Gerstein HC, Rutty CJ. Insulin therapy: The discovery that shaped a century. Canadian Journal of Diabetes. 2021;**45**(8):798-803. Available from: https://pubmed.ncbi.nlm.nih.gov/34045148/

[4] Thorens B. GLUT2, glucose sensing and glucose homeostasis. Diabetologia. 2015;**58**:221-232

[5] Atkinson MA, Campbell-Thompson M, Kusmartseva I, Kaestner KH. Organisation of the human pancreas in health and in diabetes. Diabetologia. 2020;**63**:1966-1973

[6] Tengholm A, Gylfe E. cAMP signalling in insulin and glucagon secretion. Diabetes, Obesity and Metabolism. 2017;**19**:42-53

[7] Rutter GA, Hodson DJ. Minireview: Intraislet regulation of insulin secretion in humans. Molecular Endocrinology. 2013;**27**:1984-1995

[8] Adeva-AndanyMM,Funcasta-CalderónR, Fernández-FernándezC,Castro-QuintelaE, Carneiro-Freire N. Metabolic effects of glucagon in humans. Journal of Clinical & Translational Endocrinology. 2019;**15**:45-53

[9] Hughes JW, Ustione A, Lavagnino Z, Piston DW. Regulation of islet glucagon secretion: Beyond calcium. Diabetes, Obesity and Metabolism. 2018;**20**:127-136

[10] Yu Q, Shuai H, Ahooghalandari P, Gylfe E, Tengholm A. Glucose controls glucagon secretion by directly modulating cAMP in alpha cells. Diabetologia. 2019;**62**(7):1212-1224

[11] Kyung KM, Shin HM, Jung HS, Lee EJ, Kim TK, Kim TN, et al. Comparison of pancreatic beta cells and alpha cells under hyperglycemia: Inverse coupling in pAkt-FoxO1. Diabetes Research and Clinical Practice. 2017;**131**:1-11

[12] Lee S, Dong HH. FoxO integration of insulin signaling with glucose and lipid metabolism. Journal of Endocrinology. 2017;**233**:R67-R79

[13] Tessari P. Changes in protein, carbohydrate, and fat metabolism with aging: Possible role of insulin. Nutrition Reviews. 2000;**58**(1):11-19

[14] Tessari P. Role of insulin in age-related changes in macronutrient metabolism. European Journal of Clinical Nutrition. 2000;**54**(Suppl 3):S126-S130. Available from: https://pubmed.ncbi.nlm.nih.gov/11041084/

[15] Komatsu M, Takei M, Ishii H, Sato Y. Glucose-stimulated insulin secretion: A newer perspective. Journal of Diabetes Investigation. 2013;**4**(6):511-516. Available from: https://pubmed.ncbi.nlm.nih.gov/24843702/

[16] Straub SG, Sharp GWG. Glucose-stimulated signaling pathways in biphasic insulin secretion. Diabetes/Metabolism Research and Reviews. 2002;**18**(6):451-463

[17] Wang J, Gu W, Chen C. Knocking down insulin receptor in pancreatic

beta cell lines with lentiviral-small hairpin RNA reduces glucose-stimulated insulin secretion via decreasing the gene expression of insulin, GLUT2 and Pdx1. International Journal of Molecular Sciences. 2018;**19**(4):985

[18] Poitout V, Hagman D, Stein R, Artner I, Robertson RP, Harmon JS. Regulation of the insulin gene by glucose and fatty acids. Journal of Nutrition. 2006;**136**(4):873-876 Available from: https://pubmed.ncbi.nlm.nih.gov/16549443/

[19] Tillmar L, Carlsson C, Welsh N. Control of insulin mRNA stability in rat pancreatic islets. Regulatory role of a 3′-untranslated region pyrimidine-rich sequence. The Journal of Biological Chemistry. 2002;**277**(2):1099-1106. Available from: https://pubmed.ncbi.nlm.nih.gov/11696543/

[20] Tillmar L, Welsh N. Glucose-induced binding of the polypyrimidine tract-binding protein (PTB) to the 3′-untranslated region of the insulin mRNA (ins-PRS) is inhibited by rapamycin. Molecular and Cellular Biochemistry. 2004;**260**(1):85-90

[21] Thorens B, Mueckler M. Glucose transporters in the 21st century. American Journal of Physiology. Endocrinology and Metabolism. 2010;**298**:141-145. Available from: http://www.ajpendo.org

[22] Mueckler M, Thorens B. The SLC2 (GLUT) family of membrane transporters. Molecular Aspects of Medicine. 2013;**34**(2-3):121-138

[23] Posner BI. Insulin signalling: The inside story. Can. Journal of Diabetes. 2017;**41**(1):108. Available from: /pmc/articles/PMC5272803/

[24] Newsholme EA, Dimitriadis G. Integration of biochemical and physiologic effects of insulin on glucose metabolism. Experimental and Clinical Endocrinology and Diabetes. 2001;**109**(SUPPL. 2):S122-S134. Available from: http://www.thieme-connect.de/DOI/DOI?10.1055/s-2001-18575

[25] Matschinsky FM, Wilson DF. The central role of glucokinase in glucose homeostasis: A perspective 50 years after demonstrating the presence of the enzyme in islets of Langerhans. Frontiers in Physiology. 2019;**10**:148

[26] Massa ML, Gagliardino JJ, Francini F. Liver glucokinase: An overview on the regulatorymechanisms of its activity. IUBMB Life. 2011;**63**:1-6

[27] Iynedjian PB. Molecular physiology of mammalian glucokinase. Cellular and Molecular Life Sciences. 2009;**66**: 27-42

[28] Matschinsky FM. Regulation of pancreatic-cell Glucokinase from basics to therapeutics. Diabetes. 2002;**51**(Suppl 3): S394-S404

[29] Haeusler RA, McGraw TE, Accili D. Metabolic signalling: Biochemical and cellular properties of insulin receptor signalling. Nature Reviews Molecular Cell Biology. 2018;**19**:31-44

[30] White MF. IRS proteins and the common path to diabetes. American Journal of Physiology—Endocrinology and Metabolism. 2002;**283**(3):46-43

[31] Kohn AD, Kovacina KS, Roth RA. Insulin stimulates the kinase activity of RAC-PK, a pleckstrin homology domain containing ser/thr kinase. The EMBO Journal. 1995;**14**(17):4288-4295. Available from: https://onlinelibrary.wiley.com/doi/full/10.1002/j.1460-2075.1995.tb00103.x

[32] Zorzano A, Palacín M, Gumà A. Mechanisms regulating GLUT4 glucose

transporter expression and glucose transport in skeletal muscle. Acta Physiologica Scandinavica. 2005;**183**(1):43-58

[33] Wang L, Liu Q, Kitamoto T, Hou J, Qin J, Accili D. Identification of insulin-responsive transcription factors that regulate glucose production by hepatocytes. Diabetes. 2019;**68**(6):1156-1167

[34] den Boer MAM, Voshol PJ, Kuipers F, Romijn JA, Havekes LM. Hepatic glucose production is more sensitive to insulin-mediated inhibition than hepatic VLDL-triglyceride production. American Journal of Physiology. Endocrinology and Metabolism. 2006;**291**:1360-1364. Available from: http://www.ajpendo.org

[35] Hatting M, Tavares CDJ, Sharabi K, Rines AK, Puigserver P. Insulin regulation of gluconeogenesis. Annals of the New York Academy of Sciences. 2018;**1411**:21-35

[36] Kamagate A, Dong HH. FoxO1 integrates insulin signaling to VLDL production NIH public access. Cell Cycle. 2008;**7**(20):3162-3170. Available from: http://www.landesbioscience.com/journals/cc/article/6882

[37] Kamagate A, Qu S, Perdomo G, Su D, Dae HK, Slusher S, et al. FoxO1 mediates insulin-dependent regulation of hepatic VLDL production in mice. Journal of Clinical Investigation. 2008;**118**(6):2347-2364

[38] Kim DH, Zhang T, Lee S, Calabuig-Navarro V, Yamauchi J, Piccirillo A, et al. FoxO6 integrates insulin signaling with MTP for regulating VLDL production in the liver. Endocrinology. 2014;**155**(4):1255-1267

[39] Scherer T, Ohare J, Diggs-Andrews K, Schweiger M, Cheng B, Lindtner C, et al. Brain insulin controls adipose tissue lipolysis and lipogenesis. Cell Metabolism. 2011;**13**(2):183-194

[40] Iwen KA, Scherer T, Heni M, Sayk F, Wellnitz T, Machleidt F, et al. Intranasal insulin suppresses systemic but not subcutaneous lipolysis in healthy humans. Journal of Clinical Endocrinology and Metabolism. 2014;**99**(2):E246-E251

[41] Saltiel AR. Insulin signaling in the control of glucose and lipid homeostasis. Handbook of Experimental Pharmacology. 2016;**233**:51-71. Available from: https://pubmed.ncbi.nlm.nih.gov/26721672/

[42] Charlton M, Nair KS. Protein metabolism in insulin-dependent diabetes mellitus. The Journal of Nutrition. 1998;**128**(2):323S-327S. Available from: https://academic.oup.com/jn/article/128/2/323S/4723974

[43] Fukagawa NK, Minaker KL, Rowe JW, Goodman MN, Matthews DE, Bier DM, et al. Insulin-mediated reduction of whole body protein breakdown. Dose-response effects on leucine metabolism in postabsorptive men. Journal of Clinical Investigation. 1985;**76**(6):2306-2311

[44] Newgard CB, An J, Bain JR, Muehlbauer MJ, Stevens RD, Lien LF, et al. A branched-chain amino acid-related metabolic signature that differentiates obese and lean humans and contributes to insulin resistance. Cell Metabolism. 2009;**9**(4):311-326

[45] Tessari P, Nosadini R, Trevisan R, de Kreutzenberg SV, Inchiostro S, Duner E, et al. Defective suppression by insulin of leucine-carbon appearance and oxidation in type 1, insulin-dependent

diabetes mellitus. Evidence for insulin resistance involving glucose and amino acid metabolism. Journal of Clinical Investigation. 1986;**77**(6):1797-1804

[46] Yoon MS. The emerging role of branched-chain amino acids in insulin resistance and metabolism. Nutrients. 2016;**8**(7):405

[47] Asghari G, Farhadnejad H, Teymoori F, Mirmiran P, Tohidi M, Azizi F. High dietary intake of branched-chain amino acids is associated with an increased risk of insulin resistance in adults. Journal of Diabetes. 2018;**10**(5):357-364

[48] Holst JJ, Holland W, Gromada J, Lee Y, Unger RH, Yan H, et al. Insulin and glucagon: Partners for life. Endocrinology. 2017;**158**(4):696-701

[49] Vergari E, Knudsen JG, Ramracheya R, Salehi A, Zhang Q, Adam J, et al. Insulin inhibits glucagon release by SGLT2-induced stimulation of somatostatin secretion. Nature Communications. 2019;**10**(1):139

[50] Wan CKN, Giacca A, Matsuhisa M, El-Bahrani B, Lam L, Rodgers C, et al. Increased responses of glucagon and glucose production to hypoglycemia with intraperitoneal versus subcutaneous insulin treatment. Metabolism, Clinical and Experimental. 2000;**49**(8):984-989

[51] Ito K, Maruyama H, Hirose H, Kido K, Koyama K, Kataoka K, et al. Exogenous insulin dose-dependently suppresses glucopenia-induced glucagon secretion from perfused rat pancreas. Metabolism. 1995;**44**(3):358-362

[52] Göbl C, Morettini M, Salvatori B, Alsalim W, Kahleova H, Ahrén B, et al. Temporal patterns of glucagon and its relationships with glucose and insulin following ingestion of different classes of macronutrients. Nutrients. 2022;**14**(2):376. Available from: http://www.ncbi.nlm.nih.gov/pubmed/35057557

[53] Walker JN, Ramracheya R, Zhang Q, Johnson PRV, Braun M, Rorsman P. Regulation of glucagon secretion by glucose: Paracrine, intrinsic or both? Diabetes, Obesity and Metabolism. 2011;**13**(SUPPL. 1): 95-105. Available from: https://onlinelibrary.wiley.com/doi/full/10.1111/j.1463-1326.2011.01450.x

[54] Onyango AN. Mechanisms of the regulation and dysregulation of glucagon secretion. Oxidative Medicine and Cellular Longevity. 2020;**2020**:3089139

[55] Barthel A, Schmoll D. Novel concepts in insulin regulation of hepatic gluconeogenesis. American Journal of Physiology-Endocrinology and Metabolism. 2003;**285**(4):E685-E692. Available from: http://www.ajpendo.org

[56] Cooperberg BA, Cryer PE. Insulin reciprocally regulates glucagon secretion in humans. Diabetes. 2010;**59**(11):2936-2940

[57] Bansal P, Wang Q. Insulin as a physiological modulator of glucagon secretion. American Journal of Physiology. Endocrinology and Metabolism. 2008;**295**(4):E751-E761

[58] Holst J, Christensen M, Lund A, de Heer J, Svendsen B, Kielgast U, et al. Regulation of glucagon secretion by incretins. Diabetes, Obesity and Metabolism. 2011;**13**(SUPPL. 1):89-94. Available from: https://onlinelibrary.wiley.com/doi/full/10.1111/j.1463-1326.2011.01452.x

[59] El K, Campbell JE. The role of GIP in α-cells and glucagon secretion. Peptides (NY). 2020;**125**:170213

[60] Seino Y, Fukushima M, Yabe D. GIP and GLP-1, the two incretin hormones: Similarities and differences. Journal of Diabetes Investigation. 2010;**1**:8-23

[61] Drucker DJ. Mechanisms of action and therapeutic application of glucagon-like peptide-1. Cell Metabolism. 2018;**27**(4):740-756

[62] Galicia-Garcia U, Benito-Vicente A, Jebari S, Larrea-Sebal A, Siddiqi H, Uribe KB, et al. Pathophysiology of type 2 diabetes mellitus. International Journal of Molecular Sciences. 2020;**21**:1-34

[63] Harrison LC. The dark side of insulin: A primary autoantigen and instrument of self-destruction in type 1 diabetes. Molecular Metabolism. 2021;**52**:101288

[64] Maiese K. New insights for oxidative stress and diabetes mellitus. Oxidative Medicine and Cellular Longevity. 2015;**2015**:875961

[65] Brownlee M. The pathobiology of diabetic complications a unifying mechanism. Banting Lecture. Diabetes. 2005;**54**(6):1615-1625

[66] Giacco F, Brownlee M. Oxidative stress and diabetic complications. Circulation Research. 2010;**107**:1058-1070

[67] Pi J, Zhang Q, Fu J, Woods CG, Hou Y, Corkey BE, et al. ROS signaling, oxidative stress and Nrf2 in pancreatic beta-cell function. Toxicology and Applied Pharmacology. 2010;**244**:77-83

[68] Hanley SC, Austin E, Assouline-Thomas B, Kapeluto J, Blaichman J, Moosavi M, et al. β-Cell mass dynamics and islet cell plasticity in human type 2 diabetes. Endocrinology. 2010;**151**(4):1462-1472

[69] Satin LS, Butler PC, Ha J, Sherman AS. Pulsatile insulin secretion, impaired glucose tolerance and type 2 diabetes. Molecular Aspects of Medicine. 2015;**42**:61-77

[70] Moin ASM, Dhawan S, Cory M, Butler PC, Rizza RA, Butler AE. Increased frequency of hormone negative and polyhormonal endocrine cells in lean individuals with type 2 diabetes. Journal of Clinical Endocrinology and Metabolism. 2016;**101**(10):3628-3636

[71] Roden M, Shulman GI. The integrative biology of type 2 diabetes. Nature. 2019;**576**(7785):51-60

[72] Zick Y. Insulin resistance: A phosphorylation-based uncoupling of insulin signaling. Trends in Cell Biology. 2001;**11**(11):437-441

[73] Chatterjee S, Khunti K, Davies MJ. Type 2 diabetes. The Lancet. 2017;**389**(10085):2239-2251

[74] Lewis GF, Carpentier A, Vranic M, Giacca A. Resistance to insulin's acute direct hepatic effect in suppressing steady-state glucose production in individuals with type 2 diabetes. Diabetes. 1999;**48**(3):570-576

[75] Weyer C, Bogardus C, Mott DM, Pratley RE. Introduction the natural history of insulin secretory dysfunction and insulin resistance in the pathogenesis of type 2 diabetes mellitus. Journal of Clinical Investigation. 1999;**104**(6):787-794

[76] Haeusler RA, McGraw TE, Accili D. Biochemical and cellular properties of insulin receptor signalling. Nature Reviews. Molecular Cell Biology. 2018;**19**:31-44

[77] Boucher J, Kleinridders A, Ronald KC. Insulin receptor signaling in normal and insulin-resistant states. Cold Spring Harbor Perspectives in Biology. 2014;**6**(1):a009191

[78] Kaksonen M, Roux A. Mechanisms of clathrin-mediated endocytosis. Nature Reviews. Molecular Cell Biology. 2018;**19**(5):313-326. Available from: https://pubmed.ncbi.nlm.nih.gov/29410531/

[79] Hall C, Yu H, Choi E, Insulin receptor endocytosis in the pathophysiology of insulin resistance. Experimental & Molecular Medicine. 2020;**52**(6):911-920. Available from: https://www.nature.com/articles/s12276-020-0456-3

[80] Caro JF, Ittoop O, Pories WJ, Meelheim D, Flickinger EG, Thomas F, et al. Studies on the mechanism of insulin resistance in the liver from humans with noninsulin-dependent diabetes. Insulin action and binding in isolated hepatocytes, insulin receptor structure, and kinase activity. The Journal of Clinical Investigation. 1986;**78**(1):249-258. Available from: https://pubmed.ncbi.nlm.nih.gov/3522628/

[81] Thomas DD, Corkey BE, Istfan NW, Apovian CM. Hyperinsulinemia: An early indicator of metabolic dysfunction. Journal of the Endocrine Society. 2019;**3**(9):1727. Available from: /pmc/articles/PMC6735759/

[82] Kurauti MA, Ferreira SM, Soares GM, Vettorazzi JF, Carneiro EM, Boschero AC, et al. Hyperinsulinemia is associated with increasing insulin secretion but not with decreasing insulin clearance in an age-related metabolic dysfunction mice model. Journal of Cellular Physiology. 2019;**234**(6):9802-9809. Available from: https://pubmed.ncbi.nlm.nih.gov/30370604/

[83] D'alessio D. The role of dysregulated glucagon secretion in type 2 diabetes. Diabetes, Obesity and Metabolism. 2011;**13**(SUPPL. 1):126-132

[84] Philippe J, Knepel W, Waeber G. Insulin regulation of the glucagon gene is mediated by an insulin-responsive DNA element. Proceedings of the National Academy of Sciences of the United States of America. 1991;**88**(16):7224. Available from: /pmc/articles/PMC52266/?report=abstract

[85] Eissele R, Göke R, Willemer S. Glucagon-like peptide-1 cells in the gastrointestinal tract and pancreas of rat, pig and man. European Journal of Clinical Investigation. 1992;**22**:283-291

[86] Holst JJ. The physiology of glucagon-like peptide 1. Physiological Reviews. 2007;**87**(4):1409-1439

[87] Tolhurst G, Heffron H, Lam YS, Parker HE, Habib AM, Diakogiannaki E, et al. Short-chain fatty acids stimulate glucagon-like peptide-1 secretion via the G-protein-coupled receptor FFAR2. Diabetes. 2012;**61**(2):364-371

[88] Lund A. On the role of the gut in diabetic hyperglucagonaemia. Danish Medical Journal. 2017;**64**(4):B5340

[89] Drucker DJ. Glucagon-like peptide-1 and the islet β-cell: Augmentation of cell proliferation and inhibition of apoptosis. Endocrinology. 2003;**144**:5145-5148

[90] Drucker DJ, Nauck MA. The incretin system: Glucagon-like peptide-1 receptor agonists and dipeptidyl peptidase-4 inhibitors in type 2 diabetes. Lancet. 2006;**368**(9548):1696-1705

[91] Gloyn AL, Drucker DJ. Precision medicine in the management of type 2 diabetes. The Lancet Diabetes and Endocrinology. 2018;**6**(11):891-900

[92] Lederberg J, McCray AT. Ome SweetOmics—A genealogical treasury of words. The Scientist. 2001;**15**(7):8. Available from: https://go.gale.com/ps/i.do?id=GALE%7CA73535513&sid=google

Scholar&v=2.1&it=r&linkaccess=abs&issn=08903670&p=AONE&sw=w

[93] Relman DA, Falkow S. The meaning and impact of the human genome sequence for microbiology. Trends in Microbiology. 2001;**9**(5):206-208

[94] Clavel T, Desmarchelier C, Haller D, Gérard P, Rohn S, Lepage P, et al. Intestinal microbiota in metabolic diseases: From bacterial community structure and functions to species of pathophysiological relevance. Gut Microbes. 2014;**5**(4):544-551

[95] Turnbaugh PJ, Ley RE, Hamady M, Fraser-Liggett CM, Knight R, Gordon JI. The human microbiome project. Nature. 2007;**449**(7164):804-810

[96] Clemente JC, Ursell LK, Parfrey LW, Knight R. The impact of the gut microbiota on human health: An integrative view. Cell. 2012;**148**(6):1258-1270

[97] Kau AL, Ahern PP, Griffin NW, Goodman AL, Gordon JI. Human nutrition, the gut microbiome and the immune system. Nature. 2011;**474**(7351):327-336

[98] Sato J, Kanazawa A, Ikeda F, Yoshihara T, Goto H, Abe H, et al. Gut dysbiosis and detection of "live gut bacteria" in blood of Japanese patients with type 2 diabetes. Diabetes Care. 2014;**37**(8):2343-2350

[99] Candela M, Biagi E, Soverini M, Consolandi C, Quercia S, Severgnini M, et al. Modulation of gut microbiota dysbioses in type 2 diabetic patients by macrobiotic ma-Pi 2 diet. British Journal of Nutrition. 2016;**116**(1):80-93

[100] Hooks KB, O'Malley MA. Dysbiosis and its discontents. mBio. 2017;**8**:5

[101] Carding S, Verbeke K, Vipond DT, Corfe BM, Owen LJ. Dysbiosis of the gut microbiota in disease. Microbial Ecology in Health and Disease. 2015;**26**. DOI: 10.3402/mehd.v26.26191

[102] Toor D, Wasson MK, Kumar P, Karthikeyan G, Kaushik NK, Goel C, et al. Dysbiosis disrupts gut immune homeostasis and promotes gastric diseases. International Journal of Molecular Sciences. 2019;**20**(10):2432

[103] Martinez JE, Kahana DD, Ghuman S, Wilson HP, Wilson J, Kim SCJ, et al. Unhealthy lifestyle and gut dysbiosis: A better understanding of the effects of poor diet and nicotine on the intestinal microbiome. Frontiers in Endocrinology. 2021;**12**:649

[104] Magne F, Gotteland M, Gauthier L, Zazueta A, Pesoa S, Navarrete P, et al. The firmicutes/bacteroidetes ratio: A relevant marker of gut dysbiosis in obese patients? Nutrients. 2020;**12**(5):1474

[105] Rogers GB, Keating DJ, Young RL, Wong ML, Licinio J, Wesselingh S. From gut dysbiosis to altered brain function and mental illness: Mechanisms and pathways. Molecular Psychiatry. 2016;**21**(6):738-748

[106] Bistoletti M, Caputi V, Baranzini N, Marchesi N, Filpa V, Marsilio I, et al. Antibiotic treatment-induced dysbiosis differently affects BDNF and TrkB expression in the brain and in the gut of juvenile mice. PLoS One. 2019;**14**(2):e0212856

[107] Macfarlane GT, Macfarlane S. Bacteria, colonic fermentation, and gastrointestinal health. Journal of AOAC International. 2012;**95**(1):50-60

[108] Turnbaugh PJ, Gordon JI. The core gut microbiome, energy balance

and obesity. Journal of Physiology. 2009;**587**(17):4153-4158

[109] Ley RE, Turnbaugh PJ, Klein S, Gordon JI. Microbial ecology: Human gut microbes associated with obesity. Nature. 2006;**444**(7122):1022-1023

[110] Chakraborti CK. New-found link between microbiota and obesity. World Journal of Gastrointestinal Pathophysiology. 2015;**6**(4):110

[111] Petersen C, Round JL. Defining dysbiosis and its influence on host immunity and disease. Cellular Microbiology. 2014;**16**(7):1024-1033

[112] Wei L, Singh R, Ro S, Ghoshal UC. Gut microbiota dysbiosis in functional gastrointestinal disorders: Underpinning the symptoms and pathophysiology. JGH Open. 2021;**5**(9):976-987. Available from: http://www.ncbi.nlm.nih.gov/pubmed/34584964

[113] Larsen N, Vogensen FK, van den Berg FWJ, Nielsen DS, Andreasen AS, Pedersen BK, et al. Gut microbiota in human adults with type 2 diabetes differs from non-diabetic adults. PLoS One. 2010;**5**(2):e9085

[114] Zhao L, Lou H, Peng Y, Chen S, Zhang Y, Li X. Comprehensive relationships between gut microbiome and faecal metabolome in individuals with type 2 diabetes and its complications. Endocrine. 2019;**66**(3):526-537

[115] Harsch I, Konturek P. The role of Gut microbiota in obesity and type 2 and type 1 diabetes mellitus: New insights into "old" diseases. Medical Science. 2018;**6**(2):32

[116] Kootte RS, Vrieze A, Holleman F, Dallinga-Thie GM, Zoetendal EG, de Vos WM, et al. The therapeutic potential of manipulating gut microbiota in obesity and type 2 diabetes mellitus. Diabetes, Obesity and Metabolism. 2012;**14**(2):112-120

[117] Sedighi M, Razavi S, Navab-Moghadam F, Khamseh ME, Alaei-Shahmiri F, Mehrtash A, et al. Comparison of gut microbiota in adult patients with type 2 diabetes and healthy individuals. Microbial Pathogenesis. 2017;**111**:362-369

[118] Aydin Ö, Nieuwdorp M, Gerdes V. The Gut microbiome as a target for the treatment of type 2 diabetes. Current Diabetes Reports. 2018;**18**(8):55

[119] Lambeth SM, Carson T, Lowe J, Ramaraj T, Leff JW, Luo L, et al. Composition, diversity and abundance of gut microbiome in prediabetes and type 2 diabetes. Journal of Diabetes and Obesity. 2015;**2**(3):1-7

[120] Zhao L, Zhang F, Ding X, Wu G, Lam YY, Wang X, et al. Gut bacteria selectively promoted by dietary fibers alleviate type 2 diabetes. Science (1979). 2018;**359**(6380):1151-1156

[121] Zhang F, Wang M, Yang J, Xu Q, Liang C, Chen B, et al. Response of gut microbiota in type 2 diabetes to hypoglycemic agents. Endocrine. 2019;**66**(3):485-493

[122] Mallott EK, Amato KR. Butyrate production pathway abundances are similar in human and nonhuman primate gut microbiomes. Molecular Biology and Evolution. 2022;**39**(1):msab279. Available from: https://academic.oup.com/mbe/article/39/1/msab279/6372698

[123] Nagpal R, Shively CA, Appt SA, Register TC, Michalson KT, Vitolins MZ, et al. Gut microbiome composition in non-human primates consuming a

Western or Mediterranean diet. Frontiers in Nutrition. 2018;**5**:28

[124] Clayton JB, Gomez A, Amato K, Knights D, Travis DA, Blekhman R, et al. The gut microbiome of nonhuman primates: Lessons in ecology and evolution. American Journal of Primatology. 2018;**80**(6):e22867. Available from: https://pubmed.ncbi.nlm.nih.gov/29862519/

[125] Amato KR. Incorporating the gut microbiota into models of human and non-human primate ecology and evolution. American Journal of Physical Anthropology. 2016;**159**:S196-S215

[126] Amato KR, Sanders JG, Song SJ, Nute M, Metcalf JL, Thompson LR, et al. Evolutionary trends in host physiology outweigh dietary niche in structuring primate gut microbiomes. ISME Journal. 2019;**13**(3):576-587

[127] Manara S, Asnicar F, Beghini F, Bazzani D, Cumbo F, Zolfo M, et al. Microbial genomes from non-human primate gut metagenomes expand the primate-associated bacterial tree of life with over 1000 novel species. Genome Biology. 2019;**20**(1)

[128] Wang J, Qin J, Li Y, Cai Z, Li S, Zhu J, et al. A metagenome-wide association study of gut microbiota in type 2 diabetes. Nature. 2012;**490**(7418):55-60

[129] Lazar V, Ditu LM, Pircalabioru GG, Picu A, Petcu L, Cucu N, et al. Gut microbiota, host organism, and diet trialogue in diabetes and obesity. Frontiers in Nutrition. 2019;**6**:21

[130] Meijnikman AS, Gerdes VE, Nieuwdorp M, Herrema H. Evaluating causality of gut microbiota in obesity and diabetes in humans. Endocrine Reviews. 2018;**39**(2):133-153

[131] Smits SA, Marcobal A, Higginbottom S, Sonnenburg JL, Kashyap PC. Individualized responses of gut microbiota to dietary intervention modeled in humanized mice. mSystems. 2016;**1**(5):e00098-e00116

[132] le Chatelier E, Nielsen T, Qin J, Prifti E, Hildebrand F, Falony G, et al. Richness of human gut microbiome correlates with metabolic markers. Nature. 2013;**500**(7464):541-546

[133] Gomez A, Sharma AK, Mallott EK, Petrzelkova KJ, Jost Robinson CA, Yeoman CJ, et al. Plasticity in the human gut microbiome defies evolutionary constraints. mSphere. 2019;**4**(4):e00271-e00319

[134] Wu H, Esteve E, Tremaroli V, Khan MT, Caesar R, Mannerås-Holm L, et al. Metformin alters the gut microbiome of individuals with treatment-naive type 2 diabetes, contributing to the therapeutic effects of the drug. Nature Medicine. 2017;**23**(7):850-858

[135] Tito RY, Knights D, Metcalf J, Obregon-Tito AJ, Cleeland L, Najar F, et al. Insights from characterizing extinct human Gut microbiomes. PLoS One. 2012;**7**(12):e51146

[136] Koo H, Hakim JA, Crossman DK, Kumar R, Lefkowitz EJ, Morrow CD. Individualized recovery of gut microbial strains post antibiotics. npj Biofilms and Microbiomes. 2019;**5**(1):30

[137] Murphy EF, Cotter PD, Healy S, Marques TM, O'Sullivan O, Fouhy F, et al. Composition and energy harvesting capacity of the gut microbiota: Relationship to diet, obesity and time in mouse models. Gut. 2010;**59**(12):1635-1642

[138] Ochman H, Worobey M, Kuo CH, Ndjango JBN, Peeters M, Hahn BH, et al. Evolutionary relationships of

wild hominids recapitulated by gut microbial communities. PLos Biol. 2010;**8**(11):e1000546

[139] Kushugulova A, Forslund SK, Costea PI, Kozhakhmetov S, Khassenbekova Z, Urazova M, et al. Metagenomic analysis of gut microbial communities from a central Asian population. BMJ Open. 2018;**8**(7):e021682

[140] Falony G, Joossens M, Vieira-Silva S, Wang J, Darzi Y, Faust K, et al. Population-level analysis of gut microbiome variation. Science (1979). 2016;**352**(6285):560-564

[141] Karlsson FH, Tremaroli V, Nookaew I, Bergström G, Behre CJ, Fagerberg B, et al. Gut metagenome in European women with normal, impaired and diabetic glucose control. Nature. 2013;**498**(7452):99-103

[142] Mahowald MA, Rey FE, Seedorf H, Turnbaugh PJ, Fulton RS, Wollam A, et al. Characterizing a model human gut microbiota composed of members of its two dominant bacterial phyla. Proceedings of the National academy of Sciences of the United States of America. 2009;**106**(14):5859-5864

[143] Guarner F, Malagelada JR. Gut flora in health and disease. Lancet. 2003;**361**(9356):512-519

[144] He Y, Wu W, Zheng HM, Li P, McDonald D, Sheng HF, et al. Regional variation limits applications of healthy gut microbiome reference ranges and disease models. Nature Medicine. 2018;**24**(10):1532-1535

[145] Furet JP, Kong LC, Tap J, Poitou C, Basdevant A, Bouillot JL, et al. Differential adaptation of human gut microbiota to bariatric surgery-induced weight loss: Links with metabolic and low-grade inflammation markers. Diabetes. 2010;**59**(12):3049-3057

[146] Zhang X, Shen D, Fang Z, Jie Z, Qiu X, Zhang C, et al. Human gut microbiota changes reveal the progression of glucose intolerance. PLoS One. 2013;**8**(8):e71108

[147] Koliada A, Syzenko G, Moseiko V, Budovska L, Puchkov K, Perederiy V, et al. Association between body mass index and Firmicutes/Bacteroidetes ratio in an adult Ukrainian population. BMC Microbiology. 2017;**17**(1):120

[148] Ley RE, Bäckhed F, Turnbaugh P, Lozupone CA, Knight RD, Gordon JI. Obesity alters gut microbial ecology. Proceedings of the National academy of Sciences of the United States of America. 2005;**102**(31):11070-11075

[149] Turnbaugh PJ, Hamady M, Yatsunenko T, Cantarel BL, Duncan A, Ley RE, et al. A core gut microbiome in obese and lean twins. Nature. 2008;**457**(7228):480-484. Available from: https://www.nature.com/articles/nature07540

[150] Velloso LA, Folli F, Sun XJ, White MF, Saad MJA, Kahn CR. Cross-talk between the insulin and angiotensin signaling systems. Proceedings of the National Academy of Sciences of the United States of America. 1996;**93**(22):12490-12495. Available from: https://pubmed.ncbi.nlm.nih.gov/8901609/

[151] Folli F, Kahn CR, Hansen H, Bouchie JL, Feener EP. Angiotensin II inhibits insulin signaling in aortic smooth muscle cells at multiple levels. A potential role for serine phosphorylation in insulin/angiotensin II crosstalk. The Journal of Clinical Investigation. 1997;**100**(9):2158-2169. Available from: https://pubmed.ncbi.nlm.nih.gov/9410892/

[152] Li X, Wu Y, Song Y, Ding N, Lu M, Jia L, et al. Activation of NF-κB-inducing kinase in islet β cells causes β cell failure and diabetes. Molecular Therapy. 2020;**28**(11):2430-2441

[153] Malle EK, Zammit NW, Walters SN, Koay YC, Wu J, Tan BM, et al. Nuclear factor κB-inducing kinase activation as a mechanism of pancreatic β cell failure in obesity. Journal of Experimental Medicine. 2015;**212**(8):1239-1254

[154] Karin M, Delhase M. The IκB kinase (IKK) and NF-κB: Key elements of proinflammatory signalling. Seminars in Immunology. 2000;**12**(1):85-98

[155] Meyerovich K, Fukaya M, Terra LF, Ortis F, Eizirik DL, Cardozo AK. The non-canonical NF-κB pathway is induced by cytokines in pancreatic beta cells and contributes to cell death and proinflammatory responses in vitro. Diabetologia. 2016;**59**(3):512-521

[156] Aw W, Fukuda S. Understanding the role of the gut ecosystem in diabetes mellitus. Journal of Diabetes Investigation. 2018;**9**(1):5-12. Available from: https://onlinelibrary.wiley.com/doi/full/10.1111/jdi.12673

[157] Gérard C, Vidal H. Impact of gut microbiota on host glycemic control. Frontiers in Endocrinology. 2019;**10**(JAN):29

[158] Priyadarshini M, Navarro G, Layden BT. Gut microbiota: FFAR reaching effects on islets. Endocrinology. 2018;**159**(6):2495-2505. Available from: https://pubmed.ncbi.nlm.nih.gov/29846565/

[159] Brubaker PL. Linking the gut microbiome to metabolism through endocrine hormones. Endocrinology. 2018;**159**(8):2978-2979

[160] Kamada N, Seo SU, Chen GY, Núñez G. Role of the gut microbiota in immunity and inflammatory disease. Nature Reviews. Immunology. 2013;**13**(5):321-335

[161] Maslowski KM, Vieira AT, Ng A, Kranich J, Sierro F, Yu D, et al. Regulation of inflammatory responses by gut microbiota and chemoattractant receptor GPR43. Nature. 2009;**461**(7268):1282-1286

[162] Åkerfeldt MC, Howes J, Chan JY, Stevens VA, Boubenna N, McGuire HM, et al. Cytokine-induced β-cell death is independent of endoplasmic reticulum stress signaling. Diabetes. 2008;**57**(11):3034-3044. Available from: https://diabetesjournals.org/diabetes/article/57/11/3034/13408/Cytokine-Induced-Cell-Death-Is-Independent-of

[163] Ehses JA, Böni-Schnetzler M, Faulenbach M, Donath MY. Macrophages, cytokines and β-cell death in type 2 diabetes. Biochemical Society Transactions. 2008;**36**(3):340-342

[164] Tesi M, Bugliani M, Ferri G, Suleiman M, de Luca C, Bosi E, et al. Pro-inflammatory cytokines induce insulin and glucagon double positive human islet cells that are resistant to apoptosis. Biomolecules. 2021;**11**(2):1-11

[165] Gao Z, Yin J, Zhang J, Ward RE, Martin RJ, Lefevre M, et al. Butyrate improves insulin sensitivity and increases energy expenditure in mice. Diabetes. 2009;**58**(7):1509-1517

[166] Myers MA, Hettiarachchi KD, Ludeman JP, Wilson AJ, Wilson CR, Zimmet PZ. Dietary microbial toxins and type 1 diabetes. Annals of the New York Academy of Sciences. 2003;**1005**(1):418-422. Available from: https://onlinelibrary.

wiley.com/doi/full/10.1196/annals.1288.071

[167] Cummings JH, Pomare EW, Branch HWJ, Naylor CPE, MacFarlane GT. Short chain fatty acids in human large intestine, portal, hepatic and venous blood. Gut. 1987;**28**(10):1221-1227

[168] Lin H, Frassetto A, Kowalik EJ, Nawrocki AR, Lu MM, Kosinski JR, et al. Butyrate and propionate protect against diet-induced obesity and regulate gut hormones via free fatty acid receptor 3-independent mechanisms. PLoS One. 2012;**7**(4):e35240

[169] Al-Lahham SH, Peppelenbosch MP, Roelofsen H, Vonk RJ, Venema K. Biological effects of propionic acid in humans; Metabolism, potential applications and underlying mechanisms. Biochimica et Biophysica Acta. 2010;**1801**(11):1175-1183. Available from: https://pubmed.ncbi.nlm.nih.gov/20691280/

[170] Sakakibara S, Yamauchi T, Oshima Y, Tsukamoto Y, Kadowaki T. Acetic acid activates hepatic AMPK and reduces hyperglycemia in diabetic KK-A(y) mice. Biochemical and Biophysical Research Communications. 2006;**344**(2):597-604. Available from: https://pubmed.ncbi.nlm.nih.gov/16630552/

[171] Freeland KR, Wolever TMS. Acute effects of intravenous and rectal acetate on glucagon-like peptide-1, peptide YY, ghrelin, adiponectin and tumour necrosis factor-α. British Journal of Nutrition. 2010;**103**(3):460-466. Available from: https://www.cambridge.org/core/journals/british-journal-of-nutrition/article/acute-effects-of-intravenous-and-rectal-acetate-on-glucagonlike-peptide1-peptide-yy-ghrelin-adiponectin-and-tumour-necrosis-factor/69D7F8BC56D56DB34521824AE54923B0

[172] Tang C, Offermanns S. FFA2 and FFA3 in metabolic regulation. Handbook of Experimental Pharmacology. 2017;**236**:205-220

[173] Ulven T. Short-chain free fatty acid receptors FFA2/GPR43 and FFA3/GPR41 as new potential therapeutic targets. Frontiers in Endocrinology. 2012;**3**:111

[174] Priyadarshini M, Villa SR, Fuller M, Wicksteed B, Mackay CR, Alquier T, et al. An acetate-specific GPCR, FFAR2, regulates insulin secretion. Molecular Endocrinology. 2015;**29**(7):1055-1066

[175] Remely M, Aumueller E, Merold C, Dworzak S, Hippe B, Zanner J, et al. Effects of short chain fatty acid producing bacteria on epigenetic regulation of FFAR3 in type 2 diabetes and obesity. Gene. 2014;**537**(1):85-92

[176] Priyadarshini M, Wicksteed B, Schiltz GE, Gilchrist A, Layden BT. SCFA receptors in pancreatic β cells: Novel diabetes targets? Trends in Endocrinology and Metabolism. 2016;**27**(9):653-664

[177] Li NX, Brown S, Kowalski T, Wu M, Yang L, Dai G, et al. GPR119 agonism increases glucagon secretion during insulin-induced hypoglycemia. Diabetes. 2018;**67**(7):1401-1413

[178] Mandøe MJ, Hansen KB, Hartmann B, Rehfeld JF, Holst JJ, Hansen HS. The 2-monoacylglycerol moiety of dietary fat appears to be responsible for the fat-induced release of GLP-1 in humans. The American Journal of Clinical Nutrition. 2015;**102**(3):548-555. Available from: https://pubmed.ncbi.nlm.nih.gov/26178726/

[179] Pingitore A, Chambers ES, Hill T, Maldonado IR, Liu B, Bewick G, et al. The diet-derived short chain fatty acid propionate improves beta-cell function in humans and stimulates insulin secretion from human islets in vitro. Diabetes, Obesity & Metabolism. 2017;**19**(2):257-265

[180] McNelis JC, Lee YS, Mayoral R, van der Kant R, Johnson AMF, Wollam J, et al. GPR43 potentiates β-cell function in obesity. Diabetes. 2015;**64**(9):3203-3217

[181] Wang S, Yuan M, Zhang L, Zhu K, Sheng C, Zhou F, et al. Sodium butyrate potentiates insulin secretion from rat islets at the expense of compromised expression of β cell identity genes. Cell Death and Disease. 2011;**13**(1):67. Available from: /pmc/articles/PMC8770496/

[182] Mitchell CM, Davy BM, Ponder MA, McMillan RP, Hughes MD, Hulver MW, et al. Prebiotic inulin supplementation and peripheral insulin sensitivity in adults at elevated risk for type 2 diabetes: A pilot randomized controlled trial. Nutrients. 2021;**13**(9):3235. Available from: https://pubmed.ncbi.nlm.nih.gov/34579112/

[183] Ahluwalia B, Magnusson MK, Böhn L, Störsrud S, Larsson F, Öhman L, et al. Aloe barbadensis mill. Extract improves symptoms in IBS patients with diarrhoea: Post hoc analysis of two randomized double-blind controlled studies. Therapeutic Advances in Gastroenterology. 2021;**14**:17562848211048133. Available from: https://pubmed.ncbi.nlm.nih.gov/34646359/

[184] Meyer D, Stasse-Wolthuis M. The bifidogenic effect of inulin and oligofructose and its consequences for gut health. European Journal of Clinical Nutrition. 2009;**63**(11):1277-1289. Available from: https://pubmed.ncbi.nlm.nih.gov/19690573/

[185] Kimura I, Ozawa K, Inoue D, Imamura T, Kimura K, Maeda T, et al. The gut microbiota suppresses insulin-mediated fat accumulation via the short-chain fatty acid receptor GPR43. Nature Communications. 2013;**4**:1829

[186] George Kerry R, Patra JK, Gouda S, Park Y, Shin HS, Das G. Benefaction of probiotics for human health: A review. Journal of Food and Drug Analysis. 2018;**26**(3):927-939

[187] Abenavoli L, Scarpellini E, Colica C, Boccuto L, Salehi B, Sharifi-Rad J, et al. Gut microbiota and obesity: A role for probiotics. Nutrients. 2019;**11**(11):2690

[188] Rastall RA, Gibson GR, Gill HS, Guarner F, Klaenhammer TR, Pot B, et al. Modulation of the microbial ecology of the human colon by probiotics, prebiotics and synbiotics to enhance human health: An overview of enabling science and potential applications. FEMS Microbiology Ecology. 2005;**52**(2):145-152

[189] Tao YW, Gu YL, Mao XQ, Zhang L, Pei YF. Effects of probiotics on type II diabetes mellitus: A meta-analysis. Journal of Translational Medicine. 2020;**18**(1):30

[190] Uusitalo U, Liu X, Yang J, Aronsson CA, Hummel S, Butterworth M, et al. Association of early exposure of probiotics and islet autoimmunity in the TEDDY study. JAMA Pediatrics. 2016;**170**(1):20-28

[191] Ardeshirlarijani E, Tabatabaei-Malazy O, Mohseni S, Qorbani M, Larijani B, Baradar JR. Effect of probiotics supplementation on glucose and oxidative stress in type 2 diabetes mellitus: A meta-analysis of randomized trials. DARU, Journal of Pharmaceutical Sciences. 2019;**27**(2):827-837

[192] Everard A, Lazarevic V, Derrien M, Girard M, Muccioli GM, Neyrinck AM, et al. Responses of gut microbiota and glucose and lipid metabolism to prebiotics in genetic obese and diet-induced leptin-resistant mice. Diabetes. 2011;**60**(11):2775-2786

[193] Sun Z, Sun X, Li J, Li Z, Hu Q, Li L, et al. Using probiotics for type 2 diabetes mellitus intervention: Advances, questions, and potential. Critical Reviews in Food Science and Nutrition. 2020;**60**(4):670-683

[194] Sun J, Buys NJ. Glucose- and glycaemic factor-lowering effects of probiotics on diabetes: A meta-analysis of randomised placebo-controlled trials. British Journal of Nutrition. 2016;**115**(7):1167-1177

[195] Cani PD, Knauf C, Iglesias MA, Drucker DJ, Delzenne NM, Burcelin R. Improvement of glucose tolerance and hepatic insulin sensitivity by Oligofructose requires a functional glucagon-like peptide 1 receptor. Diabetes. 2006;**55**(5):1484-1490

[196] Schmidt F, Zimmermann J, Tanna T, Farouni R, Conway T, Macpherson AJ. Noninvasive assessment of gut function using transcriptional recording sentinel cells. Science. 2022;**376**(6594):eabm6038

Chapter 4

Obesity and Gut Microbiota

Arslan Ahmad, Sakhawat Riaz and Muhammad Tanveer

Abstract

Obesity is a severe worldwide health problem driven by both hereditary and environmental factors, and its prevalence is increasing year after year. According to current thinking, The bacteria in the stomach may have a part in the growth of obesity and other health comorbidities. To better fully comprehend the link between obesity but also microbiomes, we sum up the features of the intestinal microbiota in obese people, the metabolic pathway of obesity-induced by the intestinal microbiota, and the impact of biological factors on the intestinal microbiota and adiposity in this chapter. The microbiome has been shown to have a major role in the development of obesity by regulating energy metabolism. The makeup and density of intestinal flora can be influenced by diet. Simultaneously, it is suggested that the gut microbiome be used in obesity studies. Some food items have recently shown that pro capability via functional ingredients that impact the intestinal flora, attracting the interest of scientists.

Keywords: obesity, weight loss, intestinal microbiota, diets

1. Introduction

Obesity is a physiological condition triggered by a mixture of hereditary and nongenetic variables, such as external cues. Obesity is classified by the World Health Organization as having a Body fat percentage of more than 30, however, the requirements vary by country. In China, for instance, obesity is classified as a BMI of 28 or more. Over one-third of the worldwide population is overweight, including over 10% obese, according to a thorough survey [1]. Obesity is estimated to reach 1.12 billion people worldwide by 2030 [2]. Obesity affects more than 500 million people worldwide, creating a major financial and public-health issue [3]. Obesity has sparked renewed worry and is now becoming a severe global health issue. Obesity is associated with abnormalities in triglycerides, insulin, inflammatory processes, and peroxidation, as well as a higher risk of heart disease, diabetes, and malignancy [4, 5]. According to a rising body of evidence, a bacterial imbalance in the gut contributes to obesity [6, 7]. Dietary changes, exercise, surgery, and medication are the most popular treatments for obesity. Traditional weight loss techniques, on the other hand, frequently fail to produce satisfying results, and obesity rates are expected to climb further [8]. Many dietary plants have been proven to reduce appetite, restrict food absorption, reduce adipogenesis, and increase energy consumption, and all have anti-obesity properties [9]. In the human intestinal mucosa, particularly the colon, the gut microbiota, which contains bacteria, fungi, Archea, and viruses, is common [10, 11]. The effect of gut microbiota on obesity has received a lot of attention in current history, and it might be a viable weight-loss strategy. The effects of food plants on gut flora have recently received a lot of emphasis. The gastrointestinal microbiota contains around 100 trillions of commensal bacteria, which is 10 times the body's total density [12].

To keep its birthrates high, the gut flora feeds on nutritional remnants that people cannot process, mucous secreted either by the gut, and cells waste shed as food [13]. Short-chain fats, nutrients, and right things like a pro, analgesic, or oxidative chemicals will be produced by a healthy gut bacteria, along with potentially dangerous items such as neurotoxicity, malignancies, including immunotoxins [14, 15]. Such substances can infect humans, and immediately cause mutations, thus disrupting the defense but also physiological processes of humans. As a consequence, maintaining the body's natural normal metabolic equilibrium need balanced intestinal bacteria. Obesity is regarded as a long net caloric consumption mismatch that results in excessive weight gain [16]. The interplay between biological and epigenetic variables, such as nourishment, dietary components, and/or lifestyle decisions, are to blame. Overall, the complex mechanisms that lead to obesity and its consequences are unknown, but recent research shows that it gastrointestinal tract the thousands of microorganisms that normally reside within the individual gastrointestinal tract should be taken into account [17]. Food absorption, energy management, and fat storage are all influenced by the intestinal microbiota and its microscopic genome, according to a new study.

Moreover, gut flora can alter the immune system of humans [18], and also the composition of bile salts, which can affect ingestion and physiology [19]. Obesity, cancer, and irritable bowel syndrome are hypothesized to be caused by gut microbial dysbiosis [20, 21]. Gram-negative bacteria lipopolysaccharide (LPS) may, for instance, produce an immune response in the recipient [22]. Obese people have a lesser variety of intestinal flora than lean people [23], and the huge quantity of specific gut microbiota taxa has changed in obese people [24]. Utilization of some food items could be negatively proportional to excess weight through modifying gut flora, according to epidemiological research [25–27]. As a consequence, eating dietary plants and taking advantage of their impact on gut microbiota management might be a novel method to treat obesity. These results indicate that gut bacteria may regulate the host's energy metabolism, potentially leading to obesity and other disorders. This chapter's vision is to offer a broad review of this hot issue, including the involvement of the intestinal microbiota with obesity.

2. Intestinal microbes

With around a hundred billion bacteria, the microorganisms in the human gastrointestinal system are large and varied. This colon is expected to have a bacterial cell density of 1011 to 1012 per ml, making it one of the world's most densely inhabited microbial ecosystems [28]. The gut flora contains around 3 million genetic materials and hundreds of compounds, but the genome sequence only contains approximately 23,000 genes [29]. The host intestinal flora contains 10–100 trillion germs, making it challenging for biologists to characterize the whole microbiota, particularly with the classic Sanger method.

Bacteria, fungi, and viruses are among the species that make up gut bacteria. Bacteria are divided into phylums, classes, groups, families, species, and individuals. Even though only a few phyla are included, there are over 160 species [30]. The most prevalent gut microbe phyla include Species, Acidobacteria, Lactobacillus, Lactobacillus, Actinobacteria, Microbacterium, and Verrucomicrobia, alongside

Taxa and Eubacterium [31]. accounting for 90% of a microbial population. The Firmicutes phylum has about 200 genera, including Escherichia, Staphylococcus, Coli, Enterobacter, and Ruminicoccus. The Clostridium genus makes up roughly 95% of the Firmicutes phylum. Bacteroidetes is a bacterial family that contains well-known bacteria which including Bacteroides and Prevotella. There are fewer bacteria in the Actinobacteria phylum, with the Bifidobacterium genus dominating [31].

The Firmicutes phylum, including comprises % of the gut bacteria, encompasses more than 200 genera, Escherichia, Pseudomonas, Vibrio, Enterobacter, and Ruminicoccus are a few examples. Almost the whole Genera class is represented by the Clostridium genus. Bacteroidetes is a bacterial family that includes well-known bacteria such as Bifidobacteria and Prevotella. The Lactobacillus genus dominates the Lactobacilli phylum, which contains a smaller amount of microbes.

2.1 Individual differences

Each section of the Gastrointestinal system has a different taxonomy and functional flora, which fluctuates throughout time as a result of perinatal changes, aging, and external conditions such as antimicrobial usage.

2.2 Anatomy of the intestine

Physiological factors such as acidity and high oxygen tension, digestion flow that is quick inside the lips but slows down afterward food supply, and finally human fluids all have an impact on the microbiota [32]. The gut provides a more difficult habitat for microbes due to its short transit times (3–5 hours) and high bile concentration. The largest microbial population is located in the large intestine, which has a slow mass flow and a normal to slightly acid pH, with obligate anaerobic bacteria dominating.

2.3 Evolution and resilience

Even though a tiny amount of pathogens via maternal blood could produce a first microbiome at delivery, fetuses are assumed to be sterile during pregnancy [33]. Viruses out from the mother and the surroundings infiltrate the newborn's intestines almost immediately. The microbiota's makeup is influenced by cesarean delivery, antimicrobial therapy, nutrition, and ambient hygiene [34]. Bacterial flora in your intestines is extremely stable throughout maturity, shifting just a little around a core of stable colonizers. The gut physiology and nutrition of humans alter as they get older [35]. Temporary changes, however, may occur as a result of dietary factors or antibiotic treatments. Quick medication with only a solitary prescription with antibiotic therapy, such instance, alters the intestinal flora lasting up to 4 days until returning to its previous state [36]. In addition, some bacteria might take months of rehab after treatment, resulting in a loss of diversity after multiple medication exposures [37]. Dietary modifications have a similar effect on the makeup of intestinal flora. Food provides nutrients to the host as well as the microbiota, whose bacteria may be favored or injured by dietary substrates. As a result, according to one study, changes in diet in mice could responsible for 57% of the overall point mutation in gut microbes, while gene variants accounted for only 12% [38].

So far, the gutMEGA database has collected the gut microbes of 6457 taxa [39]. Firmicutes, Bacteroides, Proteus, Actinomycetes, Fusobacteria, and Verrucomicrobia comprise the bulk of the human gut microbiota [40]. Varied gut flora is ultimately beneficial; nevertheless, a lack of choice in the gut microbiome may lead to disorders such as obesity [23]. Another feature of gut health microorganisms is a delicate balance, which refers to the gut microbiota's capacity to resist perturbation and return to health, such as following antimicrobial therapy [41].

2.4 Changes in infant and gut bacteria makeup

The mammalian gut flora is a flexible and intricate habitat that evolved including its owner [42] which accounts for around one kg of our body weight. Our intestinal bacteria populations are rapidly being recognized as an entity with physiological, immunosuppressive, and estrogen functions that lead to illnesses [43]. Each Digestive system contains around 1014 organisms ten times the level of cells in the human body and each gut flora has 500–1000 unique types of bacteria [44, 45]. The Megahits group [25] also released a list of nearly 10 million non-redundant genes derived from decoding specimens from 1267 people, showing that the microbial community includes at least 100 times the amount of genes found in the bodily genome [46–48]. An overall current population could be classified into three groups based on the nature of the gut flora [31]. The most prevalent enterotypes are Also used Prevotella, or Rotifers, with Bacteroides, Lactobacilli, and Ruminococcus leading the pack. Although enterotype differences were previously assumed to be unrelated to region, age, race, or BMI [49], they have now been connected to long-term eating habits. The gut microbiota is a symbiotic relationship that helps the human body do things it cannot. As a result, sustaining regular GI and immunological processes, as well as proper nutrition digestion, requires the gut microbiota [12, 50]. Its microbiome, for instance, ferments metabolites indigestible food elements, synthetase enzymes, and certain other critical minerals, food poisons, and carcinogens convert cholesterol and bile salts, supports immunological reaction development, controls enterocyte growth and division, controls gastrointestinal capillaries, and protects against pathogenic strains [51]. Carbohydrate composting, its generation of short-chain fatty acids, a saturation of selected surface proteins, or the formation of minerals and abundant amino acids all seem to be the main tasks of normal gut flora [52].

2.5 Gestational age at birth

Due to organ development and external influences such as antibiotics, hospitalization, and enteral feeding, colonization is a concern in preterm neonates after birth [53, 54]. For these reasons, preterm birth might have had a considerable influence on gut and systemic immunity throughout pregnancy [54].

Preterm newborns have a limited range of bacteria, with more potentially hazardous microbes again from the Proteobacteria phylum's Bacteria cell colonizing them [53] and decreased rates of strictly anaerobic bacteria like Bifidobacteria and lactobacilli [55], Bacteroides, and Atopobium [53]. Genetic factors, as well as the family's secretor and Lewis blood type, impact the composition of infant formula, resulting in four phenotypes with varied amounts of oligosaccharide [56]. Premature children born to non-secretor mothers had greater Proteobacteria levels and lower Firmicutes levels [57]. Pratic et al. [58] investigated the makeup of colostrum that discovered that Health maintenance organizations linked with different mother phenotypes influence the gut microbiota of newborns. For example, health centers associated with secretor moms might provide a prebiotic benefit by lowering microorganisms linked to sepsis and necrotizing enterocolitis [57]. This suggests that health centers can alter gut flora, protecting premature babies against gut dysfunction and NEC [59]. Lactoferrin is a well-known component of human dairy that promotes the colonizing of preterm newborns' stomachs with helpful bacteria, therefore improving their ecology [32].

2.6 Type of delivery

Babies acquire a gut that is identical to their mother's gut microflora after normal delivery. The flora of the child's large intestine and the related organisms of the vaginal tract, Bacteria, Lactobacilli, and Sneathia, were discovered to be closely linked in the development of biological baby mucus [60]. As per Biasucci et al. [61], significant bacteria such as Probiotic bacteria long and Lactobacillus catenulatum are familiar with the microbiome of perineal born neonates. *E. coli*, Staphylococcus, *Bacteroides fragilis*, and Bacteria are among the aerotolerant anaerobic bacteria found inside the infant gut [62–64].

In analyses done at 7 years old, variations in the microflora of C-section and perineal delivered infants were discovered [65]. Persistent autoimmune abnormalities like influenza, regional collagenous disorders, adolescent arthritis, irritable bowel [66], or overweight [67] have been linked to cesarean delivery.

2.7 Methods of milk feeding

As per research [68, 69], Equation babies are more likely to be contaminated with *E. coli*, Bacteroides, and Clostridium difficult than breastfeeding infants. In terms of Actinobacteria concentration, Bifidobacterium spp. has been connected to breastfeeding and artificial milk [70, 71]. In contrast to equation babies, breastfed infants have a more diverse and variable Probiotic bacteria microbiota [71]. Breastfeeding infants are provided microbiota for a more than 2 increase in Acidophilus cells as compared to supplemented infants [70]. Breastfed babies had a more favorable gut microbiota than pattern babies, with more Bifidobacterium spp. and less *Clostridium difficile* and *Escherichia coli* [23].

Maintaining a healthy and nourishing gut flora in the mother during pregnancy is also regarded to be a crucial factor in improving the milk microbiome composition. Oral supplements may increase the quantity of Acidophilus spp. and Lactobacilli spp. in human breast milk in vaginally delivered mothers [72].

2.8 Weaning period

When solid foods are introduced and dairy is eliminated, significant changes in gut flora occur. Probiotics, Escherichia coccoides, and Bifidobacteria are the most frequent species after childhood [73]. Apigenin muciniphila, Enterobacteriaceae, Veillonella, Mycobacterium coccoides spp., and Botulism spp. are all found in significant levels in the microbiota of one baby [74]. Around the age of three, the appearance and diversity of a toddler's intestinal microbiota are most akin to those of adults [75]. *Bacillus subtilis*, Bacteroidetes, and Act are the three bacterial phyla that control the adult microbial population.

2.9 Antibiotics

Pharmaceuticals could alter the intestinal microbiota's makeup to some extent. The influence of antibiotic molecular pathways on the makeup of the human microbiome was investigated in obey research [76], Penicillin treatment options alter the gut microbiota, increasing the prevalence of some species while decreasing the abundance of others. Bacterial diversity and abundance decreased during therapy. Antimicrobial class, frequency, length of therapy, pharmacokinetic properties, and target microorganisms all impact gut flora composition [77]. Antibiotic features such as antimicrobial actions and potency are important in the development of gut flora thus they are partly to blame for bacterial composition changes following antibiotic therapy [76]. The drug has unique properties and disposal methods, resulting in a wide range of bacteria material changes [77].

2.10 Gut microbiota variations between individuals

We've previously seen how single intestinal microbiota makeup changes, and now we'll examine how it varies among individuals. Intertype's, BMI levels, and extrinsic variables including behavior, health and body, race, and culinary or cultural traditions all impact cross variability.

2.11 Enterotypes

We've established that the gut flora composition differs across persons; now we'll investigate how it varies between individuals. Exogenous variables including activity regularity, race, culinary and cultural habits, enterotypes, and BMI levels all play a role in these variances. Instead of an intentional integration of germs, an enterotype is a physiologically close relation between distinct species of bacteria. Although enterotypes are not as different from plasma groups in terms of structure, they are tolerant, constant through life, and may be regained if they are changed. Enterotypes appear to be mostly defined by food habits. Knowing the genesis and roles of enterotypes might help researchers better understand the links between gut flora and people's health.

2.12 Body mass index

Many investigations [78, 79] focused on the impact of childhood obesity on intestinal flora and found that overweight or medium BMI kids had more bacterial ecology than underweight students. Intestinal flora declines with time, depending on the BMI category [78, 80]. Obese children's microbiota has a greater Firmicutes-to-Bacteroidetes ratio than lean children's microbiota, according to Bervoets et al. [81]. On the other hand, this obese microbiome exhibits comparably low percentages of Probiotic bacteria vulgatus and high levels of Escherichia species [81]. Adiposity is also linked to higher levels of Genus like Ruminococcaceae and decreased rates of Clostridium such as Bacteroidaceae and Enterobacter, according to Riva et al. [82]. Short-chain fatty acids were found to be higher in obese children, indicating that they used more fuel. Increased SCFA production and energy extraction from colon digestion are connected to a higher Firmicutes to Bacteroidetes ratio, indicating that intestinal flora imbalance might play a role in obesity etiology [82]. Gut flora instability is well predicted by BMI.

2.13 Ethnicity, dietary habits, and cultural habits

Although a healthy person's microbiome is largely stable, behavior or dietary culture choices may likely alter gut microbial behavior [49]. According to a study on European children given a Western diet and Liberia children eating a diet high in grain + local vegetables with relatively low lipids and animal protein, African children's flora contains a noticeable excess of Prevotella and Xylanibacter [83]. Shigella and *E. coli* bacteria are similarly underrepresented. Research [84] compared the intestinal microbiome of Hadza hunters with Italians. On either a phylum level, the Hadza gastrointestinal tract is dominated by Genus and Spirochaetes, whereas Cyanobacteria, a crucial upper octave member of the Italian gut microbiota, is almost non-existent. When the kind of food varies, biodegradation switches between carbohydrate and protein digestion. This occurred just one day after the food contacted the microorganisms in the distal intestine. Diet has a quick and long-term influence on the human microbiota, according to David et al. [85].

2.14 Exercise frequency

Bai et al. [78] discovered connections between exercise regularity and gut flora composition in a study of teenagers. Daily exercise increases gut microbial diversity by producing more SCFAs, via stimulating the production of adhesion molecules in colon epithelia, which may aid to improve gut barrier resilience, limiting mucosal leakage, and modulating cytokine secretion [78, 86].

3. Gut microbes in connection with obesity

The idea for studying obese people's gut microorganisms came from the idea that gut flora might be a vital component of their long-term health. The earliest evidence of a link between gut flora and obesity was discovered in germless mouse studies. The quantity of fat and insulin sensitivity inside the transplanted increased even when food consumption was reduced, showing that gut microbes may help the recipient in the formation of adipose tissue [87]. Its Firmicutes ratio rose sharply in fat mice [88], showing that the obese mice's microbiome was good at taking energy from the feed. Systems can be seen in individuals; for instance, in the guts of obese children, their ratio of Firmicutes climbed whereas the quantity of Bacteroidetes decreased [89]. The Firmicutes/Bacteroidetes ratio increased as BMI increased, according to a study of the Ukrainian population [90].

In overweight and obese people, supplementing with *A. muciniphila* improves metabolic indices [91]. Traditional probiotics like Lactobacillus and Bifidobacterium, for example, help to maintain healthy gut flora. Crovesy et al. [92] investigated the impact of Bacteria on obesity rates but found that its beneficial benefits were genus. The frequency of *Lactobacillus paracasei* was shown to be interrelated to fat, but the number of Escherichia repeating unit and *Lactobacillus acidophilus* gasseri was shown to be favorably related to obesity. Animal studies have shown that Bifidobacterium can help people lose weight. Bifidobacterium demonstrated a strain-dependent impact on obesity in diet-induced obesity animal models [93]. Obesity is linked to a reduction in Bifidobacterium abundance in the intestine [94]. The study on intestinal flora and obesity is represented in **Table 1**.

Obese and Microbe Features	Preclinical or clinical	subjects	References
Firmicutes/Bacteroidetes ratio increased	Preclinical	Mice	[12, 24]
	Clinical	Childhood	[50]
	Clinical	Adult Ukrainian population	[51]
Increased Akkermansia population reduced body weight	Clinical	Human	[52]
	preclinical	Mice	[95]
Bifidobacteria reduced	Preclinical	Rats	[55]
Methanobacteriales smithii and Bifidobacterium were associated with normal weight	Clinical	Human	[96]

Table 1.
Linkage of obesity with gut microbiomes.

3.1 Obesogenic gut microbiota

Firmicutes and Bacteroidetes, for particular, have been identified as obesity-promoting intestinal flora, which can lead to the growth of obese [97].

3.2 Firmicutes and Bacteroidetes

Ruminiococcus, Candida, and Lactobacillus have been the most prevalent representatives of the phylum Firmicutes phylum Bacteroidetes in the gut bacteria, accounting for 90% of types of bacteria [44, 98]. Regulating glucagon-like peptide 1 release may aid to alleviate insulin sensitivity and obesity in way of eating obese C57BL/6 J mice given antibiotics [99]. Inside the intestines of adult C57BL/6 J rats fed a strong diet, firmicutes were found mainly [100]. In obese people and obese mice, a great proportion of Firmicutes to Bacteroidetes has just been reported as an adiposity trait of the gut microbiome [42]. Obese women having elevated toll-like receptor 5 gene expression were also shown to have a greater number of the genus [101]. Egyptian researchers examined the gut microbiome of 51 obese persons (23 kids and 28 individuals) to the gut microbiome of 28 healthy individuals in a study. In a survey of 17 children and 11 adults, researchers observed that the phyla Firmicutes and Bacteroidetes were significantly higher in the obese group (p 0.001, $p = 0.003$) [102]. Lactobacillus has been divided into various subgroups, each of which has been associated with obesity and the genesis of obesity. A variety of key enzymes are missing in bacteria that promote weight gain, including sugar enzymes, antioxidants, and dextrin, L-rhamnose, or acetate synthetases [103].

The three principal Bacteroidetes taxa present in the human stomach are Bacteroides, Prevotella, and Porphyromonas. Bacteroides account for more than a third of all gut bacteria, and it's particularly prevalent in Westerners who consume a high-fat or high-sugar diet [104]. Together in a controlled trial with 138 babies aged 3 years, the utilization of Bacteroides in the intestines was found to be positively associated with bodyweight [105]. Bifidobacteria and Lactobacillus species have also been linked to weight increase in children [106, 107].

3.3 Anti-obesity gut microbiota

Certain gut microbiota species have been reported to have anti-obesity characteristics, in contrast to the obesogenic gut microbiota. In the next part, Bifid bacteria, Lactobacillus subspecies, and Bacteroidetes are investigated as anti-obesity gut microbiota.

3.4 Probiotics and obesity

C57BL/6 J mice were given *Bifidobacterium lactis* 420 for 12 weeks to inhibit weight gain, which may be attributed to decreased intestinal epithelial adhesion and blood LPS [108]. Probiotic lactic 420 also improved the viability and lowered the porosity of Overexpressing cells in a dose-dependent manner, suggesting that it might help with the treatment of low-grade inflammatory disorders like obesity [109]. During eight weeks, High fat-feed mice were given Bacteria bacillus bifidum BGN4 and Probiotics reticulata BORI, which significantly reduced weight gain and lowered liver triglycerides and total cholesterol, as well as blood aspartate and alanine transaminase activity [110].

Despite the reality that some Lacto strains were linked to obesity, most of the Bacillus species were found to have a pro function [98]. Lactobacillus aided fat loss in animals, whereas Bacterium gasseri aided weight loss for both obese people and animals, as per a meta-analysis [111]. Lactobacillus cultures 031 CE reduced lipid levels and the activity of aspartate transaminase and alanine transaminase in the hepatic Institute of Cancer Research mice high-fat- fat diet [110]. With down-regulating-regulating TNF-, interleukin-1, and Nuclear factor and upregulating IL-10 and tight junction, Bacteria sakei OK67 treatment to fat-fed mice greatly lower body or epididymides fat excess weight [112]. Par-, PR domain containing 16, Par- coactivator-1, growth factor protein 7, and fibroblast growth factor 21, were all increased by Bacteria consisting of a resistor 263 in Adult male rats [113].

4. Obesity mechanism induced by gut micro-biota

4.1 Energy absorption

To provide energy to their humans, obese rats consume more carbohydrates via their gut bacteria [114, 115]. When bacteria mice colonized predominantly by the obesity microbiota' did not change their food or weight, their total body fat increased in comparison to mice colonized by the Chilean biome' [88]. Obese people's gut microbiota has a larger capacity for absorbing energy from meals, according to the study. Obese mice had higher lipid uptake, according to a multi-omics study. In germ-free mice, Clostridia colonization downregulates genetic variants' fat intake [47]. In a way, the gut bacteria of fat people may produce more impact energy, leading to higher energy and weight growth. Difficult-to-digest carbohydrates are fermented by gut bacteria into short-chain fatty acids, which are either eaten or expelled in the stool. Short-chain fatty acids are necessary to maintain energy balance [116]. SCFAs have lately received a lot of attention for their positive effects on cellular integrity and lipid metabolism, although their relevance in obesity is still debatable. Intestinal permeability, metabolic disease indicators, obesity, and hypertension have all been associated with increased fecal SCFA levels [117].

4.2 Central appetite and fat accumulation

The gut microbiota has become one of the most transcription factors of intestine connection. Within the study of the morphological and molecular origins of obesity and associated illnesses, the gastrointestinal system pathway has attracted much interest. Hormonal, immunological, or neurological pathways connect both brains with the microbiome [118]. The intestinal microbiota link influences the nerve cells of said individual. The autonomic nervous system can affect the makeup and structure of the gut flora. Microbiomes affect cognitive activity in a variety of ways, including by influencing the synthesis of neuropeptides like dopamine, which are critical for gastrointestinal function regulation [119]. Lactate, a nerve terminal fuel generated by Lactobacillus and Bifidobacterium, has now been demonstrated to enhance satisfaction following a meal [120]. Protracted hunger suppression controlled via hypothalamic neurotransmitter energetic pathways can be paired with short-term stomach pleasure regulation linked with bacterial proliferation [121]. **Table 2** depicts the obesity process as part of gut flora.

In 2004, it was shown that gut microbes can influence fat accumulation [87]. The gut bacteria upregulates two key signaling pathways, glycemic reaction component binding domain or cholesterol control component related proteins, causing fat to accumulate in hepatic. Lipids directly stimulate through the liver, where they can be absorbed via visceral fat, thanks to lipoprotein lipase. Fiaf, an LPL regulator, is produced by intestinal epithelial cells. Normal mouse intestinal epithelial cells have Fiaf inhibited, allowing the host to store more energy.

Influence	Features of Microbes	Process	References
Load capacity has increased.	Streptomyces depletion with Sulfolobus proliferation	The expression in genes that govern lipid absorption, such as CD36, has increased.	[122]
The host will have more energy.	Fusobacterium, Roseburia feces, and other Cycle life grew in number, whereas *Akkermansia muciniphila*, *Alistipes finegoldii*, Bacteroides, Christensenellaceae, Methanobrevibacter, and Oscillospira dropped in the count.	Short-chain fatty acids in abundance	[117]
Hunger rise	Clostridial clusters XIVa and IV prevail in this colony.	Neuropeptide levels were significantly lower in obese subjects.	[123–125]
Fat accumulation has increased.	Gut bacteria from normally grown mice were transferred into microbe mice.	Increased synthesis of Articulate and Depositors, which activates LPL and helps triglyceride entry into the bloodstream out from the liver, suppresses Fiaf.	[87]

Table 2.
Adiposity caused by gut microbes.

5. Obesity and microbiota: Connectivity to genetic makeup and transport

A combination of genetic and chemical variables impacts obesity. The microbe is thought to be influenced by inheritance. In actuality, several gene mutations might be responsible for changes in the structure but also diversity of the intestinal microbiota in obese people. A connection between twin genetic variation and distinct microbial species was discovered using whole-genome correlation. More than a dozen gut microbes have been linked to good health [126]. Genes affect bacteria, as evidenced by Probiotic bacteria and the lactose intolerance genome cluster [126] and AMY1-CN as candidate genes linked to the shape and severity of the microbial [127]. It's also possible that the gut microbiota is handed down from mother to kid. The gastrointestinal tract of spore mice was shown to be relatively stable in succession studies. In most cases, these bacteria make up a great proportion of the gut flora of mice, suggesting that rodents get the majority of their intestinal flora from their mothers [128]. The microbial community may be detected in the womb, synovial fluid, amniotic fluid, and even mucus, according to the study, so parental microbes could have a major impact on the development of the child's microflora [129]. Obesity is caused by a variety of causes, one of which is a bad diet. In industrialized nations and places, the consumption of high-fat and high-sugar meals has steadily increased, increasing obesity. Changes in nutrition have a profound impact on intestinal flora since gut bacteria rely on human food for survival and energy. Bacteroidetes were detected in reduced quantities in rats given a strong diet, although Firmicutes or Proteobacteria were found in higher levels. Similar changes were observed in mice who were not overweight, implying as saturated cholesterol would have a detrimental influence on this microbiota [130, 131]. The gut microbiota can be dysbiotic due to both hereditary and environmental causes. **Figure 1** shows that dysregulation could indeed affect energy uptake through transcriptional but also heavily rely on short-chain lipids, and also enhance core hunger via the intestine pivot, intestines estrogen, or neuropeptides; restrict fat metabolism via signaling pathways and glycoprotein lysozyme; trigger serious swelling via immunomodulation cell proliferation but also lipoteichoic acid, and obstruct the sleep cycle by influencing. Obesity vulnerability tends to be enhanced by these variables.

Sleep deprivation can also contribute to obesity. Sleep deprivation can impact intestinal flora and thus cause weight gain by interrupting sleep cycles. Insomnia led to huge dietary intake and long-term alterations within gut flora, with Lactobacillaceae and Ruminococcus content levels increasing and Lactobacillaceae abundance values dropping. These factors promote peripheral and visceral white adipose tissue irritation, and glycemic control changes [132]. Stress stimulates desire that leads to overweight by the application that regulates metabolism thus promoting the ingestion of desserts and fats meals [133].

5.1 Eating flora pro impact via modifying gut flora

Fruits, veggies, peppers, cereals, grain, and tea are just a few of the foods that were demonstrated to reduce obesity through modulating the microbiota and activity in the intestine [9, 134, 135].

5.2 Fruits

Fruit is rich in phenolics, pectin, and xylose, which may help prevent obesity, cancer, and heart disease [136, 137]. A 122-person randomized trial in the United

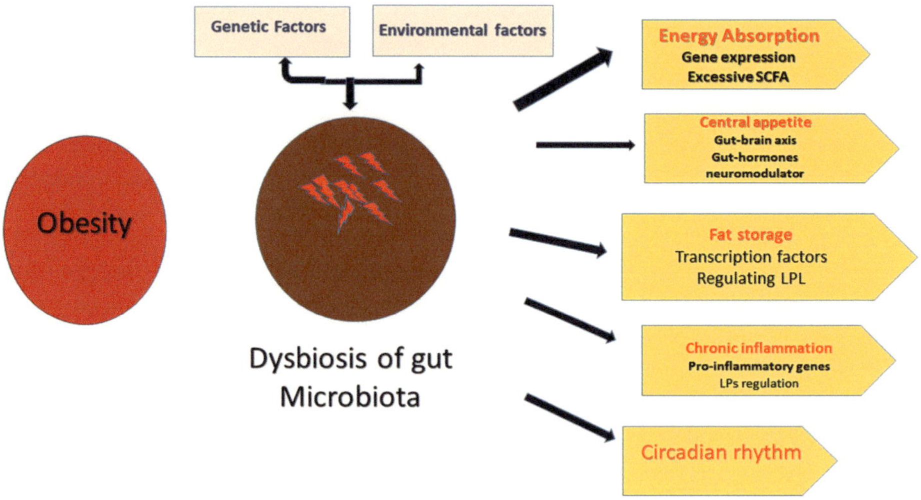

Figure 1.
Obesity emerges as a result of the microbiota's dysregulation, which is produced by the microbiota's immediate touch with local cells.

Kingdom discovered that increasing fruit and vegetable consumption altered the aspects of the human intestinal flora, with an increase through Clostridium providing a broad range of bromine and a decrease in pathogens Clostridia, which could be linked to obesity prevention [138].

As shown in **Figure 2**, taking pro supplements Organic vegetables in the diet boosts pro gut flora while reducing obesity-promoting bacteria. Toxins produced by a gastrointestinal microbiota with amplitude modulation may help with weight loss by lowering ghrelin, reducing fat storage by bottom triglycerides and up-regulating adipocytes charring, enhancing gastric mucosal feature, but also reducing intestine soreness by lowering Tumor necrosis factor, Nuclear factor, or Lipid polysaccharides, and improving gut mucosal function.

5.3 Grapes

Grapes are a nutrient-dense fruit that is abundant in resveratrol, a natural flavonoid with a slew of health advantages [136, 137]. In HFD-fed rats, resveratrol reduced weight growth and subcutaneous adipose weight while boosting the Bacteroidetes to Genera ratios, Streptococcus, and Probiotics while decreasing *E. coli* faecalis. The anti-obesity effects of resveratrol may be due to lower gene expression of medical field enzymes such as lipolysis, acylated deaminase 1, propyl hydroxylase 1, or fatty acid synthase [134]. In contrast, feeding C57BL/6 J mice grape pomace and cinnamon bark extract for 8 weeks lowered obesity by lowering fat mass, adipose irritation, and modifying gut microbiota and intestinal barrier indicators. Allobaculum and Rose bury were up-regulated in C57BL/6 J mice following treatment with combined extracts, Enzyme activities, and Lactobacillus, on the other hand, were away back [139].

5.4 Apples

Firmicutes, Bifidobacteria, *E. coli*, Enterobacter, Vibrio cholera, but also Probiotics were brewed with fecal matter from nutrition obese mice, and the crude extract was

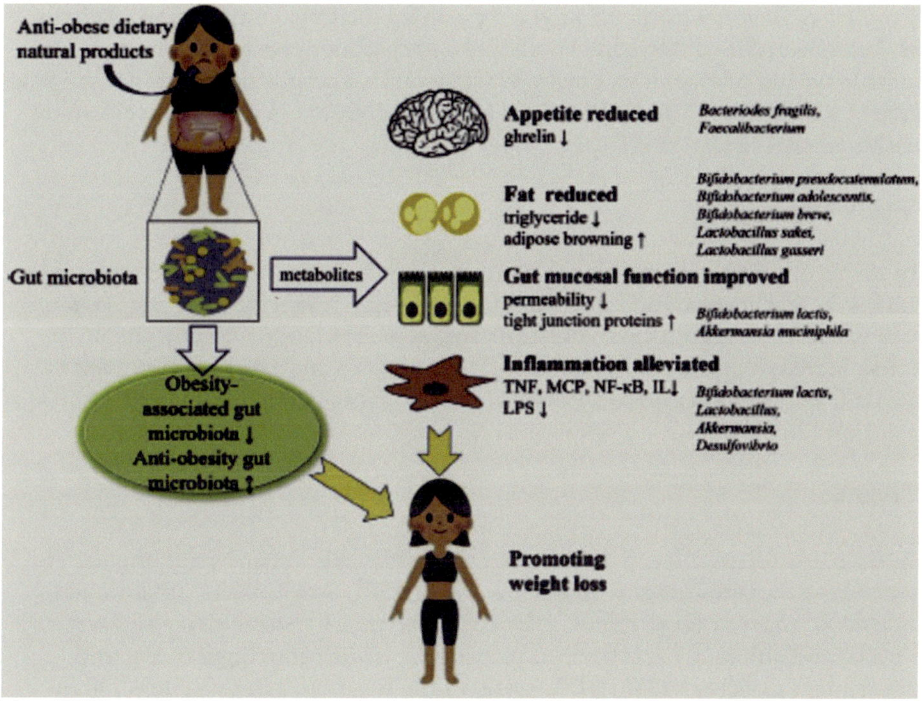

Figure 2.
Mechanisms of dietary plant weight reduction benefits through gut bacteria modifications.

able to control the gut flora of organisms linked to obesity by altering the volumes of Genus, Bifidobacteria, Enterobacter, *E.coli*, *Escherichia coli* Moreover, 0.5% polymerization semi fruit fulfillment reduced budget deficit obesity in rats fed a high-fat by lowering the Genera to Bacteroidetes ratios while eightfold doubling the levels of Akkermansia [140].

5.5 Berries

Many currants, such as blueberries, black currants, and plant-feeding-feeding-feeding, have been known as the anti properties by affect the gut flora [141–143]. By lowering Tumor necrosis factor or interleukin levels, enhancing insulin production, and raising Gammaproteobacteria density in Wistar rats, blueberries powdered may protect them against High fat-induced inflammation [144].

Pro bacteria like Akkermansia and Desulfovibrio can also be increased and whole black raspberries may lower intestinal inflammation [145]. Proanthocyanidins, a polyphenols duo prevalent in strawberries, were given to adult Zealand white bunnies for twelve weeks to alleviate nutrition adiposity by raising the quantity of Bifidobacteria at the phylum level and Akkermansia at the genera [146].

5.6 Other fruits

In diabetic mice on the High - fat diet, mangoes with 10% restored the frequency of Bifidobacteria and Akkermansia, lowering intestinal microbiota coccidiosis [147].

Poly methoxy flavones and hydroxyl poly ethoxy flavones, both found in citrus peels, have been shown to reduce body mass and adipocytes bulk in high-fat-fed mice by lowering oil droplets, perilipin 1 nutrients, and glycosides controlling signal sequence 1, as well as raising Prevotella and reducing rc4–4 microbes within rat digestive tract [148].

5.7 Vegetables

In terms of attributes, a chloroplast component found in all green vegetable tissue has been shown to enhance weight reduction in rats by boosting *Bacteroides fragilis* while boosting hunger [149]. In cross-sectional research with healthy females, increasing soluble fiber intake from veggies and fruits was proven to reduce tall weight gain and increase Ruminococcaceae abundance, and improved respiration [26].

5.8 Legumes

In nutrition obese mice, pea flour had a considerable anti-obesity impact and enhanced the Bacteroidetes to Firmicutes ratio [150]. Soy proteins are known to reduce rat fat mass or fat percentage by 10%, enhance hepatotoxicity and tertiary bile acids, and enhance Lactobacillus prevalence while reducing Blautia, and Lachnospiraceae richness [151]. Likewise, mung bean proteins, that are high in 8-globulin, are said to reduce adiposity formation and excess weight caused by the HFD, as well as ketoacidosis [152]. Mung bean proteins, on the other hand, were linked to an increase in Bifidobacteria and just a decline in Genus, a raise in intestine glucosidase potential associated tiers, and a higher primary biliary total acidity.

5.9 Tea

Tea has been a popular beverage for a long time. Tea has recently shown anti-obesity capabilities through a variety of means, including lowering fat accumulation in cells and changing gut microbiota [153]. However, dosing of C57BL/6 J mice with crude extract of green, oolong, and black tea indicated that these tea extracts improved glycemic control and also decreased weight gain through modifying the gut microbiota. The Rikenellaceae and Desulfovibrionaceae families were decreased in number, leading to greater SCFA levels, lower lipopolysaccharide tiers, and improved glucose metabolism [154]. By enhancing the percentages of Genus to Bifidobacteria and Bifidobacteria to Lactobacilli, and also flattening the looks of lipid metabolism and offensive genetic makeup in white adipose tissue, kefir black tea effectively reduced weight gain but also abdominal obesity in obese rats without any influence on caloric intake [155].

5.10 Spices, turmeric and chili

Herbs have such a longstanding experience of usage in food flavoring, while polyphenol found in spices has been demonstrated to get a variety of bioactivity, Anti-obesity, anti-cancer, anti-inflammatory, and anti-bacterial growth suppression are only a few of the benefits [156, 157]. Turmeric contains curcumin, which is a key bioactive component with a lengthy range of health benefits. Turmeric has been demonstrated to have a significant effect on the public of certain intestinal microbiota in mice, notably Lactobacilli, Bacteroidaceae, and Rikenella, which have both been linked to obesity-related illnesses [135]. Curcumin decreased weight gain in obese menopausal rats without affecting estrogen levels and improved gut microbiota diversity [158].

Because capsaicin is a key component of chili's bioactive components, it's one of the most popular hot flavors. According to studies [159], capsaicin reduced weight gain and inactivated the muscarinic receptor type 1 in rats on the High - fat diet. Capsaicin reduced microorganisms and increased Aeruginosa muciniphila in high-fat-fed mice [160].

5.11 Obesity and short-chain fatty acids

The most prevalent compounds in gut flora are sterols, which have some important pharmacological roles in keeping the host alive. By functioning as a link between the intestinal microbiota and the host, these chemicals influence barrier function, irritation regulation, bile salt conversion, immunological activity, and infection control. Despite their modest levels in the vascular, acetate and benzoate have direct impacts on organs by activating the hormonal and neurologic systems. For example, pectin is both a fossil fuel for epithelium and a histone deacetylase inhibitor that affects gene expression and cell destiny [161]. In adults, phytic acid suppresses fatty acid synthesis while simultaneously acting as a moderate pro in the gut [162]. Likewise, the microbiota's citric acid serves a variety of physiologic purposes. It is a precursor for lipid production [163], and an appetite suppressant via a primary hypothalamus pathway [164].

Short-chain fatty acids, primarily butyric acid, provide around 70% of fuel to the epithelium [165]. Acetic, propionic, and butyric acids can thus operate as both anabolic nutrients and chemical messengers in a wide range of cell activities [34].

Indigestible carbohydrate fibfibersres provide an extra biochemical energy source for the gut flora. Sulfonamides, the major metabolic byproducts, can be used for Vivo lipid or glycogen production [12]. The change in Short-chain- chain fatty acids levels in obese could be attributed to intestinal bacteria in the gut microbiome. This complex microbial population has a higher metabolic ability and performs a variety of tasks in the human gut [87].

The gut bacteria aid in the breakdown of raw carbohydrates into readily digestible oligosaccharides, as well as villus epithelial triglyceride lipase activity and Short-chain fatty acids synthesis [166], both of which are important for the host's nutrition and energy management. Intestinal bacteria may contribute to obesity by increasing nutrition and altering host lipid metabolic activity, as well as fueling homeostasis through its metabolites [167]. It's not unexpected that changes in intestinal flora diversity, with Firmicutes being more numerous in obese than lean patients, cause problems with energy uptake and management [28]. In the Netherlands, obese and overweight people exhibited higher fecal matter Short-chain- chain fatty acids concentrations and more Genera than their slim equivalents, according to research. Obese people are expected to yield more colon SCFA, implying a higher microbial power harvest [168, 169], confirming the theory that changes in SCFA levels in obesity are caused by dysregulation in the colon microbiome. Even more clearly, the gut flora influences weight control via SCFAs, altering energy imbalance and DNA synthesis through miRNAs [79].

6. Conclusion

Every person's gut microbiota is unique to them. In the formative years (4–36 months), intestine maturity shapes fundamental native flora, which is influenced by that does, birth gestation age, type of delivery, milky nursing techniques, weaning duration, lifestyle, and dietary and sociocultural practices. The gut microbiota, which plays a vital role in individual energy balance, is connected to obesity. Because some gut microorganisms associated with Lactobacillus, Genera, and Bifidobacteria are linked to weight increase, whereas Bifidobacterium, most Lactobacillus, and some Bifidobacteria have anti-obesity functions, the effects of intestinal microbiota on obesity development are species-dependent. Obesity is linked to a dysregulation of gut flora. Obesity has indeed been connected to a variety of bacteria in the intestine. They raise the recipient's elastic modulus, and hypothalamic desire, including fat deposition, promoting the start and progression of obesity. Because of the diversity and variety of gut microbes, the strategy whereby it induces obesity needs to be researched further. Adiposity is the outcome of the interaction of genetic and environmental factors. A range of dietary items, including fruits (grapes, apples, and berries), vegetables, spices, legumes, cereals, and tea, have been demonstrated to modify the composition in some recent experimental and epidemiological investigations. Obese and overweight persons have greater amounts of Short-chain-chain fatty acids and more Genera in their feces than slender ones. Future research will concentrate on research methodology using survey strategy to best investigate the function of the intestinal flora link, substitute research of conservation concerns to spot possible microbial delegates of gut bacteria related to diet, and specific microbiomes regulation for obese people.

Acknowledgements

We thank the digital library GCUF for providing access to the publication.

Conflict of interest

There is no conflict of interest as declared by all authors.

Author details

Arslan Ahmad[1*], Sakhawat Riaz[1] and Muhammad Tanveer[2]

1 Department of Home Economics, Government College University Faisalabad, Pakistan

2 Department of Nutritional Sciences, Government College University Faisalabad, Pakistan

*Address all correspondence to: ahmadarslan784@gmail.com

References

[1] GBD 2015 Obesity Collaborators, Afshin A, Forouzanfar MH, Reitsma MB, Sur P, Estep K, et al. Health effects of overweight and obesity in 195 countries over 25 years. The New England Journal of Medicine. 2017;**377**:13-27 [PMID: 28604169. DOI: 10.1056/NEJMoa1614362

[2] Kelly T, Yang W, Chen CS, Reynolds K, He J. Global burden of obesity in 2005 and projections to 2030. International Journal of Obesity. 2008;**32**:1431-1437 [PMID: 18607383. DOI: 10.1038/ijo.2008.102

[3] Swinburn BA, Sacks G, Hall KD, McPherson K, Finegood DT, Moodie ML, et al. The global obesity pandemic: Shaped by global drivers and local environments. Lancet. 2011;**378**(9793):804-814. DOI: 10.1016/S0140-6736(11)60813-1

[4] Saltiel AR, Olefsky JM. Inflammatory mechanisms linking obesity and metabolic disease. The Journal of Clinical Investigation. 2017;**127**:1-4 [PMID: 28045402. DOI: 10.1172/JCI92035

[5] Bendor CD, Bardugo A, Pinhas-Hamiel O, Afek A, Twig G. Cardiovascular morbidity, diabetes, and cancer risk among children and adolescents with severe obesity. Cardiovascular Diabetology. 2020;**19**:79 [PMID: 32534575. DOI: 10.1186/s12933-020-01052-1

[6] Sonnenburg JL, Bäckhed F. Diet-microbiota interactions as moderators of human metabolism. Nature. 2016;**535**:56-64 [PMID: 27383980. DOI: 10.1038/nature18846

[7] Gomes AC, Hoffmann C, Mota JF. The human gut microbiota: Metabolism and perspective in obesity. Gut Microbes. 2018;**9**:308-325 [PMID: 29667480. DOI: 10.1080/19490976.2018.1465157

[8] Kazemipoor M, Cordell GA, Sarker MMR, Radzi CWJBWM, Hajifaraji M, Kiat PE. Alternative treatments for weight loss: Safety/risks and effectiveness of anti-obesity medicinal plants. International Journal of Food Properties. 2015;**18**(9):1942-1963

[9] Fu CH, Jiang Y, Guo J, Su ZQ. Natural products with anti-obesity effects and different mechanisms of action. Journal of Agricultural and Food Chemistry. 2016;**64**:9571-9585

[10] Qin J, Li R, Raes J, Arumugam M, Burgdorf KS, Manichanh C, et al. A human gut microbial gene catalog established by metagenomic sequencing. Nature. 2010;**464**:59e65

[11] Lopez-Legarrea P, Fuller NR, Zulet MA, Martinez JA, Caterson ID. The influence of Mediterranean, carbohydrate, and high protein diets on gut microbiota composition in the treatment of obesity and associated inflammatory state. Asia Pacific Journal of Clinical Nutrition. 2014;**23**:360e8

[12] Bäckhed F, Ley RE, Sonnenburg JL, Peterson DA, Gordon JI. Host-bacterial mutualism in the human intestine. Science. 2005;**307**:1915-1920 [PMID: 15790844. DOI: 10.1126/science.1104816

[13] Gentile CL, Weir TL. The gut microbiota at the intersection of diet and human health. Science. 2018;**362**:776-780 [PMID: 30442802. DOI: 10.1126/science.aau5812

[14] Miyamoto J, Igarashi M, Watanabe K, Karaki SI, Mukouyama H, Kishino S, et al. Gut microbiota confers host

resistance to obesity by metabolizing dietary polyunsaturated fatty acids. Nature Communications. 2019;**10**:4007 [PMID: 31488836. DOI: 10.1038/s41467-019-11978-0

[15] Canfora EE, Meex RCR, Venema K, Blaak EE. Gut microbial metabolites in obesity, NAFLD, and T2DM. Nature Reviews. Endocrinology. 2019;**15**:261-273 [PMID: 30670819. DOI: 10.1038/s41574-019-0156-z

[16] Hill JO. Understanding and addressing the epidemic of obesity: An energy balance perspective. Endocrine Reviews. 2006;**27**(7):750-761

[17] Clavel T, Desmarchelier C, Haller D, Gerard P, Rohn S, Lepage P, et al. Intestinal microbiota in metabolic diseases: From bacterial community structure and functions to species of pathophysiological relevance. Gut Microbes. 2014;**5**(4):544-551. DOI: 10.4161/gmic.29331

[18] Round JL, Mazmanian SK. The gut microbiota shapes intestinal immune responses during health and disease. Nature Reviews Immunology. 2009;**9**:313-323

[19] Sayin SI, Wahlstrom A, Felin J, Jantti S, Marschall HU, Bamberg K, et al. Gut microbiota regulates bile acid metabolism by reducing the levels of tauro-β-muricholic acid, a naturally occurring FXR antagonist. Cell Metabolism. 2013;**17**:225-235

[20] Gilbert JA, Blaser MJ, Caporaso JG, Jansson JK, Lynch SV, Knight R. Current understanding of the human microbiome. Nature Medicine. 2018;**24**(4):392-400

[21] Postler TS, Ghosh S. Understanding the holobiont: How microbial metabolites affect human health and shape the immune system. Cell Metabolism. 2017;**26**(1):110-130

[22] Shen J, Obin MS, Zhao L. The gut microbiota, obesity, and insulin resistance. Molecular Aspects of Medicine. 2013;**34**:39-58

[23] Turnbaugh PJ, Hamady M, Yatsunenko T, Cantarel BL, Duncan A, Ley RE, et al. A core gut microbiome in obese and lean twins. Nature. 2009;**457**:480-484

[24] Ley RE, Bäckhed F, Turnbaugh P, Lozupone CA, Knight RD, Gordon JI. Obesity alters gut microbial ecology. Proceedings of the National Academy of Sciences of the United States of America. 2005;**102**:11070-11075

[25] Kang C, Zhang Y, Zhu XH, Liu K, Wang XL, Chen MT, et al. Healthy subjects differentially respond to dietary capsaicin correlating with specific gut enterotypes. Journal of Clinical Endocrinology & Metabolism. 2016;**101**(12):4681-4689

[26] Menni C, Jackson MA, Pallister T, Steves CJ, Spector TD, Valdes AM. Gut microbiome diversity and high fiber intake are related to lower long-term weight gain. International Journal of Obesity. 2017;**41**(7):1099-1105

[27] Moreno-Indias I, Sánchez-Alcoholado L, Pérez-Martinez P, Andrés-Lacueva C, Cardona F, Tinahones F, et al. Red wine polyphenols modulate fecal microbiota and reduce markers of the metabolic syndrome in obese patients. Food & Function. 2016;**7**(4):1775-1787

[28] Ley RE, Turnbaugh PJ, Klein S, Gordon JI. Microbial ecology: Human gut microbes associated with obesity. Nature. 2006;**444**:1022-1023

[29] Valdes AM, Walter J, Segal E, Spector TD. Role of the gut microbiota in nutrition and health. BMJ. 2018;**361**:36-44

[30] Laterza L, Rizzatti G, Gaetani E, Chiusolo P, Gasbarrini A. The gut microbiota and immune system relationship in human graft-versus-host disease. Mediterranean Journal of Hematology Infections Diseases. 2016;**8**:e2016025

[31] Arumugam M, Raes J, Pelletier E, Le Paslier D, Yamada T, Mende DR, et al. Enterotypes of the human gut microbiome. Nature. 2011;**473**: 174-180

[32] Mastromarino P, Capobianco D, Campagna G, Laforgia N, Drimaco P, Dileone A, et al. Correlation between lactoferrin and beneficial microbiota in breast milk and infant's feces. Biometals. 2014;**27**:1077-1086

[33] Aagaard K, Ma J, Antony KM, Ganu R, Petrosino J, Versalovic J. The placenta harbors a unique microbiome. Science Translational Medicine. 2014;**6**(237):237ra265. DOI: 10.1126/scitranslmed.3008599

[34] Palmer C, Bik EM, DiGiulio DB, Relman DA, Brown PO. Development of the human infant intestinal microbiota. PLoS Biology. 2007;**5**(7):e177

[35] Faith JJ, Guruge JL, Charbonneau M, Subramanian S, Seedorf H, Goodman AL, et al. The long-term stability of the human gut microbiota. Science. 2013;**341**(6141):1237439. DOI: 10.1126/science.1237439

[36] Dethlefsen L, Huse S, Sogin ML, Relman DA. The pervasive effects of an antibiotic on the human gut microbiota, as revealed by deep 16S rRNA sequencing. PLoS Biology. 2008;**6**(11):e280. DOI: 10.1371/journal.pbio.0060280

[37] Dethlefsen L, Relman DA. Incomplete recovery and individualized responses of the human distal gut microbiota to repeated antibiotic perturbation. Proceedings National Academic Science USA. 2011;**108** (Suppl 1):4554-4561. DOI: 10.1073/pnas.1000087107

[38] Zhang C, Zhang M, Wang S, Han R, Cao Y, Hua W, et al. Interactions between gut microbiota, host genetics and diet relevant to the development of metabolic syndromes in mice. The ISME Journal. 2010;**4**(2):232-241. DOI: 10.1038/ismej.2009.112

[39] Zhang Q, Yu K, Li S, et al. gutMEGA: A database of the human gut MEtaGenome Atlas. Briefings in Bioinformatics. May 2021;**22**(3):bbaa082. DOI: 10.1093/bib/bbaa082. PMID: 32496513

[40] Consortium HMP. Structure, function, and diversity of the healthy human microbiome. Nature. 2012;**486**:207-214 [PMID: 22699609. DOI: 10.1038/nature11234

[41] Bäckhed F, Fraser CM, Ringel Y, Sanders ME, Sartor RB, Sherman PM, et al. Defining a healthy human gut microbiome: Current concepts, future directions, and clinical applications. Cell Host & Microbe. 2012;**12**:611-622 [PMID: 23159051. DOI: 10.1016/j.chom.2012.10.012

[42] Ley RE, Hamady M, Lozupone C, Turnbaugh PJ, Ramey RR, Bircher JS, et al. Evolution of mammals and their gut microbes. Science. 2008;**320**(5883):1647-1651

[43] O'Hara AM, Shanahan F. The gut flora as a forgotten organ. EMBO

Reports. 2006;**7**(7):688-693. DOI: 10.1038/sj.embor.7400731

[44] Eckburg PB, Bik EM, Bernstein CN, Purdom E, Dethlefsen L, Sargent M, et al. Diversity of the human intestinal microbial flora. Science. 2005;**308**(5728):1635-1638

[45] Gill SR, Pop M, Deboy RT, Eckburg PB, Turnbaugh PJ, Samuel BS, et al. Metagenomic analysis of the human distal gut microbiome. Science. 2006;**312**(5778):1355-1359

[46] Li J, Jia H, Cai X, Zhong H, Feng Q, Sunagawa S, et al. An integrated catalog of reference genes in the human gut microbiome. Nature Biotechnology. 2014;**32**(8):834-841. DOI: 10.1038/nbt.2942

[47] Dethlefsen L, McFall-Ngai M, Relman DA. An ecological and evolutionary perspective on human-microbe mutualism and disease. Nature. 2007;**449**(7164):811-818

[48] Qin J, Li R, Raes J, Arumugam M, Burgdorf KS, Manichanh C, et al. A human gut microbial gene catalogue established by metagenomic sequencing. Nature. 2010;**464**(7285):59-65

[49] Wu GD, Chen J, Hoffmann C, Bittinger K, Chen YY, Keilbaugh SA, et al. Linking long-term dietary patterns with gut microbial enterotypes. Science. 2011;**334**(6052):105-108. DOI: 10.1126/science.1208344

[50] Hooper LV, Midtvedt T, Gordon JI. How host-microbial interactions shape the nutrient environment of the mammalian intestine. Annual Review of Nutrition. 2002;**22**:283-307

[51] Gerard P, Bernalier-Donadille A. Les fonctions majeures du microbiote intestinal. Cahiers de Nutrition et de Die'te'tique. 2007;**42**:S28-S36

[52] Lloyd-Price J, Abu-Ali G, Huttenhower C. The healthy human microbiome. Genome Medicine. 2016;**8**:51 [PMID: 27122046. DOI: 10.1186/s13073-016-0307-y

[53] Arboleya S, Binetti A, Salazar N, Fernández N, Solís G, Hernández-Barranco A, et al. Establishment and development of intestinal microbiota in preterm neonates. FEMS Microbiology Ecology. 2012;**79**:763-772

[54] Ren S, Hui Y, Obelitz-Ryom K, Brandt AB, Kot W, Nielsen DS, et al. Neonatal gut and immune maturation is determined more by postnatal age than by post-conceptional age in moderately preterm pigs. American Journal of Physiology. Gastrointestinal and Liver Physiology. 2018;**315**:G855-G867

[55] Butel MJ, Suau A, Campeotto F, MagneF, AiresJ, FerrarisL, etal. Conditions of bifidobacterial colonization in preterm infants: A prospective analysis. Journal of Pediatric Gastroenterology and Nutrition. 2007;**44**:577-582

[56] Gabrielli O, Zampini L, Galeazzi T, Padella L, Santoro L, Peila C, et al. Preterm milk oligosaccharides during the first month of lactation. Pediatrics. 2011;**128**:e1520-e1531

[57] Underwood MA, Gaerlan S, De Leoz ML, Dimapasoc L, Kalanetra KM, Lemay DG, et al. Human milk oligosaccharides in premature infants: Absorption, excretion, and influence on the intestinal microbiota. Pediatric Research. 2015;**78**:670-677

[58] Praticò G, Capuani G, Tomassini A, Baldassarre ME, Delfini M, Miccheli A. Exploring human breast milk composition by NMR-based metabolomics. Natural Product Research. 2014;**28**:95-101

[59] Bering SB. Human milk oligosaccharides to prevent gut dysfunction and necrotizing enterocolitis in preterm neonates. Nutrients. 2018;**10**:1461

[60] Dominguez-Bello MG, Costello EK, Contreras M, Magris M, Hidalgo G, Fierer N, et al. Delivery mode shapes the acquisition and structure of the initial microbiota across multiple body habitats in newborns. Proceedings National Academic Science USA. 2010;**107**:11971-11975

[61] Biasucci G, Benenati B, Morelli L, Bessi E, Boehm G. Cesarean delivery may affect the early biodiversity of intestinal bacteria. The Journal of Nutrition. 2008;**138**:1796S-1800S

[62] Mueller NT, Bakacs E, Combellick J, Grigoryan Z, Dominguez-Bello MG. The infant microbiome development: Mom matters. Trends in Molecular Medicine. 2015;**21**:109-117

[63] Pantoja-Feliciano IG, Clemente JC, Costello EK, Perez ME, Blaser MJ, Knight R, et al. Biphasic assembly of the murine intestinal microbiota during early development. The ISME Journal. 2013;**7**:1112-1115

[64] Grönlund MM, Lehtonen OP, Eerola E, Kero P. Fecal microflora in healthy infants born by different methods of delivery: Permanent changes in intestinal flora after cesarean delivery. Journal of Pediatric Gastroenterology and Nutrition. 1999;**28**:19-25

[65] Salminen S, Gibson GR, McCartney AL, Isolauri E. Influence of mode of delivery on gut microbiota composition in seven year old children. Gut. 2004;**53**:1388-1389

[66] Sevelsted A, Stokholm J, Bønnelykke K, Bisgaard H. Cesarean section and chronic immune disorders. Pediatrics. 2015;**135**:e92-e98

[67] Chen G, Chiang WL, Shu BC, Guo YL, Chiou ST, Chiang TL. Associations of caesarean delivery and the occurrence of neurodevelopmental disorders, asthma or obesity in childhood based on Taiwan birth cohort study. BMJ Open. 2017;**7**:e017086

[68] Azad MB, Konya T, Maughan H, Guttman DS, Field CJ, Chari RS, et al. Gut microbiota of healthy Canadian infants: Profiles by mode of delivery and infant diet at 4 months. CMAJ. 2013;**185**:385-394

[69] Penders J, Thijs C, Vink C, Stelma FF, Snijders B, Kummeling I, et al. Factors influencing the composition of the intestinal microbiota in early infancy. Pediatrics. 2006;**118**:511-521

[70] Bezirtzoglou E, Tsiotsias A, Welling GW. Microbiota profile in feces of breast- and formula-fed newborns by using fluorescence in situ hybridization (FISH). Anaerobe. 2011;**17**:478-482

[71] Roger LC, Costabile A, Holland DT, Hoyles L, McCartney AL. Examination of faecal Bifidobacterium populations in breast- and formula-fed infants during the first 18 months of life. Microbiology. 2010;**156**:3329-3341

[72] Mastromarino P, Capobianco D, Miccheli A, Praticò G, Campagna G, Laforgia N, et al. Administration of a multistrain probiotic product (VSL#3) to women in the perinatal period differentially affects breast milk beneficial microbiota in relation to mode of delivery. Pharmacological Research. 2015;**95**:63-70

[73] Fallani M, Amarri S, Uusijarvi A, Adam R, Khanna S, Aguilera M, et al. Determinants of the human infant

intestinal microbiota after the introduction of first complementary foods in infant samples from five European centres. Microbiology. 2011;**157**:1385-1392

[74] Tidjani Alou M, Lagier JC, Raoult D. Diet influence on the gut microbiota and dysbiosis related to nutritional disorders. Human Microbiome Journal. 2016;**1**:3-11

[75] Yatsunenko T, Rey FE, Manary MJ, Trehan I, Dominguez-Bello MG, Contreras M, et al. Human gut microbiome viewed across age and geography. Nature. 2012;**486**:222-227

[76] Pérez-Cobas AE, Artacho A, Knecht H, Ferrús ML, Friedrichs A, Ott SJ, et al. Differential effects of antibiotic therapy on the structure and function of human gut microbiota. PLoS One. 2013;**8**:e80201

[77] Iizumi T, Battaglia T, Ruiz V, Perez GI. Gut microbiome and antibiotics. Archives of Medical Research. 2017;**48**:727-734

[78] Bai J, Hu Y, Bruner DW. Composition of Gut Microbiota and its Association with Body Mass Index and Lifestyle Factors in a Cohort of 7-18 Years Old Children from the American Gut Project. Obes: Pediatr; 2018

[79] Karlsson CL, Onnerfält J, Xu J, Molin G, Ahrné S, Thorngren-Jerneck K. The microbiota of the gut in preschool children with normal and excessive body weight. Obesity. 2012;**20**:2257-2261

[80] Yun Y, Kim HN, Kim SE, Heo SG, Chang Y, Ryu S, et al. Comparative analysis of gut microbiota associated with body mass index in a large Korean cohort. BMC Microbiology. 2017;**17**:151

[81] Bervoets L, Van Hoorenbeeck K, Kortleven I, Van Noten C, Hens N, Vael C, et al. Differences in gut microbiota composition between obese and lean children: A cross-sectional study. Gut Pathogens. 2013;**5**:10

[82] Riva A, Borgo F, Lassandro C, Verduci E, Morace G, Borghi E, et al. Pediatric obesity is associated with an altered gut microbiota and discordant shifts in Firmicutes populations. Environmental Microbiology. 2017;**19**:95-105

[83] De Filippo C, Cavalieri D, Di Paola M, Ramazzotti M, Poullet JB, Massart S, et al. Impact of diet in shaping gut microbiota revealed by a comparative study in children from Europe and rural Africa. Proceeding of National Academic Science USA. 2010;**107**:14691-14696

[84] Schnorr SL, Candela M, Rampelli S, Centanni M, Consolandi C, Basaglia G, et al. Gut microbiome of the Hadza hunter-gatherers. Nature Communications. 2014;**5**:3654

[85] David LA, Maurice CF, Carmody RN, Gootenberg DB, Button JE, Wolfe BE, et al. Diet rapidly and reproducibly alters the human gut microbiome. Nature. 2014;**505**:559-563

[86] Monda V, Villano I, Messina A, Valenzano A, Esposito T, Moscatelli F, et al. Exercise modifies the gut microbiota with positive health effects. Oxidative Medicine and Cellular Longevity. 2017;**3831972**

[87] Bäckhed F, Ding H, Wang T, Hooper LV, Koh GY, Nagy A, et al. The gut microbiota as an environmental factor that regulates fat storage. Proceeding of National Academic Science USA. 2004;**101**:15718-15723 [PMID: 15505215. DOI: 10.1073/pnas.0407076101

[88] Turnbaugh PJ, Ley RE, Mahowald MA, Magrini V, Mardis ER,

Gordon JI. An obesity-associated gut microbiome with increased capacity for energy harvest. Nature. 2006;**444**:1027-1031 [PMID: 17183312. DOI: 10.1038/nature05414

[89] Indiani CMDSP, Rizzardi KF, Castelo PM, Ferraz LFC, Darrieux M, Parisotto TM. Childhood obesity and Firmicutes/Bacteroidetes ratio in the gut microbiota: A systematic review. Childhood Obesity. 2018;**14**:501-509 [PMID: 30183336. DOI: 10.1089/chi.2018.0040

[90] Koliada A, Syzenko G, Moseiko V, Budovska L, Puchkov K, Perederiy V, et al. Association between body mass index and Firmicutes/Bacteroidetes ratio in an adult Ukrainian population. BMC Microbiology. 2017;**17**:120 [PMID: 28532414. DOI: 10.1186/s12866-017-1027-1

[91] Depommier C, Everard A, Druart C, Plovier H, Van Hul M, Vieira-Silva S, et al. Supplementation with Akkermansia muciniphila in overweight and obese human volunteers: A proofof-concept exploratory study. Nature Medicine. 2019;**25**:1096-1103 [PMID: 31263284. DOI: 10.1038/s41591-019-0495-2

[92] Crovesy L, Ostrowski M, Ferreira DMTP, Rosado EL, Soares-Mota M. Effect of lactobacillus on body weight and body fat in overweight subjects: A systematic review of randomized controlled clinical trials. International Journal of Obesity. 2017;**41**:1607-1614 [PMID: 28792488. DOI: 10.1038/ijo.2017.161

[93] Yin YN, Yu QF, Fu N, Liu XW, Lu FG. Effects of four Bifidobacteria on obesity in high-fat diet-induced rats. World Journal of Gastroenterology. 2010;**16**:3394-3401 [PMID: 20632441. DOI: 10.3748/wjg.v16.i27.3394

[94] Waldram A, Holmes E, Wang Y, Rantalainen M, Wilson ID, Tuohy KM, et al. Top-down systems biology modeling of host metabotype-microbiome associations in obese rodents. Journal of Proteome Research. 2009;**8**:2361-2375 [PMID: 19275195. DOI: 10.1021/pr8009885

[95] Anhê FF, Roy D, Pilon G, Dudonné S, Matamoros S, Varin TV, et al. A polyphenol-rich cranberry extract protects from diet-induced obesity, insulin resistance and intestinal inflammation in association with increased Akkermansia spp. population in the gut microbiota of mice. Gut. 2015;**64**:872-883 [PMID: 25080446. DOI: 10.1136/gutjnl-2014-307142

[96] Million M, Maraninchi M, Henry M, Armougom F, Richet H, Carrieri P, et al. Obesity-associated gut microbiota is enriched in lactobacillus reuteri and depleted in Bifidobacterium animalis and Methanobrevibacter smithii. International Journal of Obesity. 2012;**36**:817-825 [PMID: 21829158. DOI: 10.1038/ijo.2011.153

[97] Prior RL, Wilkes S, Rogers T, Khanal RC, Wu X, Hager TJ, et al. Dietary black raspberry anthocyanins do not alter the development of obesity in mice fed an obesogenic high-fat diet. Journal of Agricultural and Food Chemistry. 2010;**58**:3977-3983

[98] Barlow GM, Yu A, Mathur R. Role of the gut microbiome in obesity and diabetes mellitus. Nutrition in Clinical Practice. 2015;**30**:787-797

[99] Hwang I, Park YJ, Kim YR, Kim YN, Ka S, Lee HY, et al. Alteration of gut microbiota by vancomycin and bacitracin improves insulin resistance via glucagon-like peptide 1 in diet-induced obesity. Federation of American Societies

for Experimental Biology Journal. 2015;**29**(6):2397-2411

[100] Shang Y, Khafipour E, Derakhshani H, Sarna LK, Woo CW, Siow YL, et al. A short-term high-fat diet induces obesity-enhancing changes in mouse gut microbiota that are partially reversed by cessation of the high-fat diet. Lipids. 2017;**52**(6):499-511

[101] Pekkala S, Munukka E, Kong L, Pollanen E, Autio R, Roos C, et al. Toll-like receptor 5 in obesity: The role of gut microbiota and adipose tissue inflammation. Obesity. 2015;**23**(3):581-590

[102] Ismail NA, Ragab SH, Abd ElBaky A, Shoeib ARS, Alhosary Y, Fekry D. Frequency of Firmicutes and Bacteroidetes in gut microbiota in obese and normal weight Egyptian children and adults. Archives of Medical Science. 2011;**7**(3):501-507

[103] Drissi F, Merhej V, Angelakis E, El Kaoutari A, Carrière F, Henrissat B, et al. Comparative genomics analysis of lactobacillus species associated with weight gain or weight protection. Nutrition & Diabetes. 2014;**4**:e109

[104] Johnson EL, Heaver SL, Walters WA, Ley RE. Microbiome and metabolic disease: Revisiting the bacterial phylum Bacteroidetes. Journal of Molecular Medicine. 2017;**95**(1):1-8

[105] Vael C, Verhulst SL, Nelen V, Goossens H, Desager KN. Intestinal microflora and body mass index during the first three years of life: An observational study. Gut Pathogens. 2011;**3**:8

[106] Koleva PT, Bridgman SL, Kozyrskyj AL. The infant gut microbiome: Evidence for obesity risk and dietary intervention. Nutrients. 2015;**7**(4):2237-2260

[107] Kolida S, Gibson GR. Synbiotics in health and disease. Annual Review of Food Science and Technology. 2011;**2**:373-393

[108] Stenman LK, Waget A, Garret C, Klopp P, Burcelin R, Lahtinen S. Potential probiotic Bifidobacterium animalis ssp. lactis 420 prevents weight gain and glucose intolerance in diet-induced obese mice. Beneficial Microbes. 2014;**5**(4):437-445

[109] Mokkala K, Laitinen K, Röytiö H. Bifidobacterium lactis 420 and fish oil enhance intestinal epithelial integrity in Caco-2 cells. Nutrition Research. 2016;**36**(3):246-252

[110] Li Z, Jin H, Oh SY, Ji GE. Anti-obese effects of two lactobacilli and two Bifidobacteria on ICR mice fed on a high-fat diet. Biochemical and Biophysical Research Communications. 2016;**480**(2):222-227

[111] Million M, Angelakis E, Paul M, Armougom F, Leibovici L, Raoult D. Comparative meta-analysis of the effect of lactobacillus species on weight gain in humans and animals. Microbial Pathogenesis. 2012;**53**(2):100-108

[112] Lim SM, Jeong JJ, Woo KH, Han MJ, Kim DH. Lactobacillus sakei OK67 ameliorates high-fat diet-induced blood glucose intolerance and obesity in mice by inhibiting gut microbiota lipopolysaccharide production and inducing colon tight junction protein expression. Nutrition Research. 2016;**36**(4):337-348

[113] Chen LH, Chen YH, Cheng KC, Chien TY, Chan CH, Tsao SP, et al. Antiobesity effect of lactobacillus reuteri

263 associated with energy metabolism remodeling of white adipose tissue in high-energy-diet-fed rats. Journal of Nutritional Biochemistry. 2018;**54**:87-94

[114] Krajmalnik-Brown R, Ilhan ZE, Kang DW, DiBaise JK. Effects of gut microbes on nutrient absorption and energy regulation. Nutrition in Clinical Practice. 2012;**27**:201-214 [PMID: 22367888. DOI: 10.1177/08845336 11436116

[115] Festi D, Schiumerini R, Eusebi LH, Marasco G, Taddia M, Colecchia A. Gut microbiota and metabolic syndrome. World Journal of Gastroenterology. 2014;**20**:16079-16094 [PMID: 25473159. DOI: 10.3748/wjg.v20.i43.16079

[116] Cani PD, Van Hul M, Lefort C, Depommier C, Rastelli M, Everard A. Microbial regulation of organismal energy homeostasis. Nature Metabolism. 2019;**1**:34-46 [PMID: 32694818. DOI: 10.1038/s42255-018-0017-4

[117] de la Cuesta-Zuluaga J, Mueller NT, Álvarez-Quintero R, Velásquez-Mejía EP, Sierra JA, Corrales Agudelo V, et al. Higher fecal short-chain fatty acid levels are associated with gut microbiome dysbiosis, obesity, hypertension and cardiometabolic disease risk factors. Nutrients. 2018;**11** [PMID: 30591685. DOI: 10.3390/ nu11010051

[118] Cryan JF, O'Riordan KJ, CSM C, Sandhu KV, TFS B, Boehme M, et al. The microbiota-gut-brain axis. Physiological Reviews. 2019;**99**:1877-2013 [PMID: 31460832. DOI: 10.1152/ physrev.00018.2018

[119] Shajib MS, Khan WI. The role of serotonin and its receptors in the activation of immune responses and inflammation. Acta Physiologica (Oxford, England). 2015;**213**:561-574 [PMID: 25439045. DOI: 10.1111/ apha.12430

[120] Silberbauer CJ, Surina-Baumgartner DM, Arnold M, Langhans W. Prandial lactate infusion inhibits spontaneous feeding in rats. American Journal of Physiology. Regulatory, Integrative and Comparative Physiology. 2000;**278**:R646-R653 [PMID: 10712284. DOI: 10.1152/ ajpregu.2000.278.3.R646

[121] Fetissov SO. Role of the gut microbiota in host appetite control: Bacterial growth to animal feeding behavior. Nature Reviews. Endocrinology. 2017;**13**:11-25 [PMID: 27616451. DOI: 10.1038/nrendo. 2016.150

[122] Petersen C, Bell R, Klag KA, Lee SH, Soto R, Ghazaryan A, et al. T cell-mediated regulation of the microbiota protects against obesity. Science. 2019;**365** [PMID: 31346040. DOI: 10.1126/science.aat9351

[123] Wu Y, He H, Cheng Z, Bai Y, Ma X. The role of neuropeptide Y and peptide YY in the development of obesity via the gut-brain axis. Current Protein & Peptide Science. 2019;**20**:750-758 [PMID: 30678628. DOI: 10.2174/13892037206661 90125105401

[124] Salehi M, Purnell JQ. The role of glucagon-like peptide-1 in energy homeostasis. Metabolic Syndrome and Related Disorders. 2019;**17**:183-191 [PMID: 30720393. DOI: 10.1089/ met.2018.0088

[125] Federico A, Dallio M, Tolone S, Gravina AG, Patrone V, Romano M, et al. Gastrointestinal hormones, intestinal microbiota and metabolic homeostasis in obese patients: Effect of bariatric surgery. In Vivo. 2016;**30**: 321-330 [PMID: 27107092]

[126] Goodrich JK, Davenport ER, Beaumont M, Jackson MA, Knight R, Ober C, et al. Genetic determinants of the gut microbiome in UK twins. Cell Host & Microbe. 2016;**19**:731-743 [PMID: 27173935. DOI: 10.1016/j.chom.2016.04.017

[127] Poole AC, Goodrich JK, Youngblut ND, Luque GG, Ruaud A, Sutter JL, et al. Human salivary amylase gene copy number impacts oral and gut microbiomes. Cell Host & Microbe. 2019;**25**:553-564.e7 [PMID: 30974084. DOI: 10.1016/j.chom.2019.03.001

[128] Moeller AH, Suzuki TA, Phifer-Rixey M, Nachman MW. Transmission modes of the mammalian gut microbiota. Science. 2018;**362**:453-457 [PMID: 30361372. DOI: 10.1126/science.aat7164

[129] Koleva PT, Kim JS, Scott JA, Kozyrskyj AL. Microbial programming of health and disease starts during fetal life. Birth Defects Research. Part C, Embryo Today. 2015;**105**:265-277 [PMID: 26663884. DOI: 10.1002/bdrc.21117

[130] Rothschild D, Weissbrod O, Barkan E, Kurilshikov A, Korem T, Zeevi D, et al. Environment dominates over host genetics in shaping the human gut microbiota. Nature. 2018;**555**:210-215 [PMID: 29489753. DOI: 10.1038/nature25973

[131] Liu MT, Huang YJ, Zhang TY, Tan LB, Lu XF, Qin J. Lingguizhugan decoction attenuates dietinduced obesity and hepatosteatosis via the gut microbiota. World Journal of Gastroenterology. 2019;**25**:3590-3606 [PMID: 31367159. DOI: 10.3748/wjg.v25.i27.3590

[132] Poroyko VA, Carreras A, Khalyfa A, Khalyfa AA, Leone V, Peris E, et al. Chronic sleep disruption alters gut microbiota, induces systemic and adipose tissue inflammation and insulin resistance in mice. Scientific Reports. 2016;**6**:35405 [PMID: 27739530. DOI: 10.1038/srep35405

[133] Kuperman Y, Issler O, Regev L, Musseri I, Navon I, Neufeld-Cohen A, et al. Perifornical Urocortin-3 mediates the link between stress-induced anxiety and energy homeostasis. Proceedings of National Academic Science USA. 2010;**107**:8393-8398 [PMID: 20404164. DOI: 10.1073/pnas.1003969107

[134] Qiao Y, Sun J, Xia S, Tang X, Shi Y, Le G. Effects of resveratrol on gut microbiota and fat storage in a mouse model with high-fat-induced obesity. Food & Function. 2014b;**5**(6):1241-1249

[135] Shen L, Liu L, Ji HF. Regulative effects of curcumin spice administration on gut microbiota and its pharmacological implications. Food & Nutrition Research. 2017a;**61**:1361780

[136] Zhang YJ, Li S, Gan RY, Zhou T, Xu DP, Li HB. Impacts of gut bacteria on human health and diseases. International Journal of Molecular Sciences. 2015a;**16**:7493-7519

[137] Zhang YJ, Gan RY, Li S, Zhou Y, Li AN, Xu DP, et al. Antioxidant phytochemicals for the prevention and treatment of chronic diseases. Molecules. 2015b;**20**:21138-21156

[138] Klinder A, Shen Q, Heppel S, Lovegrove JA, Rowland I, Tuohy KM. Impact of increasing fruit and vegetables and flavonoid intake on the human gut microbiota. Food & Function. 2016;**7**(4):1788-1796

[139] Van Hul M, Geurts L, Plovier H, Druart C, Everard A, Ståhlman M, et al. Reduced obesity, diabetes, and steatosis upon cinnamon and grape pomace are

associated with changes in gut microbiota and markers of gut barrier. American Journal of Physiology - Endocrinology and Metabolism. 2018;**314**(4):334-352

[140] Condezo-Hoyos L, Mohanty IP, Noratto GD. Assessing non-digestible compounds in apple cultivars and their potential as modulators of obese faecal microbiota in vitro. Food Chemistry. 2014;**161**:208-215

[141] Jiang TT, Gao XJ, Wu C, Tian F, Lei QC, Bi JC, et al. Apple-derived pectin modulates gut microbiota, improves gut barrier function, and attenuates metabolic endotoxemia in rats with diet-induced obesity. Nutrients. 2016;**8**(3):126. DOI: 10.3390/nu 8030126

[142] Masumoto S, Terao A, Yamamoto Y, Mukai T, Miura T, Shoji T. Nonabsorbable apple procyanidins prevent obesity associated with gut microbial and metabolomic changes. Scientific Reports. 2016;**6**:31208

[143] Heyman-Lindén L, Kotowska D, Sand E, Bjursell M, Plaza M, Turner C, et al. Lingonberries alter the gut microbiota and prevent low-grade inflammation in high-fat diet fed mice. Food & Nutrition Research. 2016;**60**:29993

[144] Lee S, Keirsey KI, Kirkland R, Grunewald ZI, Fischer JG, de La Serre CB. Blueberry supplementation influences the gut microbiota, inflammation, and insulin resistance in high-fat-diet-fed rats. Journal of Nutrition. 2018b;**148**(2):209-219

[145] Pan P, Lam V, Salzman N, Huang YW, Yu J, Zhang J, et al. Black raspberries and their anthocyanin and fiber fractions alter the composition and diversity of gut microbiota in F-344 rats. Nutrition and Cancer-An International Journal. 2017;**69**(6):943-951

[146] Xing YW, Lei GT, Wu QH, Jiang Y, Huang MX. Procyanidin B2 protects against diet-induced obesity and nonalcoholic fatty liver disease via the modulation of the gut microbiota in rabbits. World Journal of Gastroenterology. 2019;**25**:955-966

[147] Ojo B, El-Rassi G, Payton ME, Perkins-Veazie P, Clarke S, Smith BJ, et al. Mango supplementation modulates gut microbial dysbiosis and short-chain fatty acid production independent of body weight reduction in C57BL/6 mice fed a high-fat diet. Journal of Nutrition. 2016;**146**(8):1483-1491

[148] Tung YC, Chang WT, Li S, Wu JC, Badmeav V, Ho CT, et al. Citrus peel extracts attenuated obesity and modulated gut microbiota in mice with high-fat diet-induced obesity. Food & Function. 2018;**9**(6):3363-3373

[149] Stenblom EL, Weström B, Linninge C, Bonn P, Farrell M, Rehfeld JF, et al. Dietary green-plant thylakoids decrease gastric emptying and gut transit, promote changes in the gut microbial flora, but does not cause steatorrhea. Nutrition & Metabolism. 2016;**13**:67

[150] Eslinger AJ, Eller LK, Reimer RA. Yellow pea fiber improves glycemia and reduces Clostridium leptum in diet-induced obese rats. Nutrition Research. 2014;**34**(8):714-722

[151] Panasevich MR, Schuster CM, Phillips KE, Meers GM, Chintapalli SV, Wankhade UD, et al. Soy compared with milk protein in a Western diet changes fecal microbiota and decreases hepatic steatosis in obese OLETF rats. Journal of Nutritional Biochemistry. 2017;**46**:125-136

[152] Nakatani A, Li X, Miyamoto J, Igarashi M, Watanabe H, Sutou A, et al.

Dietary mung bean protein reduces high-fat diet-induced weight gain by modulating host bile acid metabolism in a gut microbiota-dependent manner. Biochemical and Biophysical Research Communications. 2018;**501**(4):955-961

[153] Pan MH, Tung YC, Yang G, Li S, Ho CT. Molecular mechanisms of the anti-obesity effect of bioactive compounds in tea and coffee. Food & Function. 2016;**7**(11):4481-4491

[154] Liu JH, Hao WJ, He ZY, Kwek E, Zhao YM, Zhu HY, et al. Beneficial effects of tea water extracts on the body weight and gut microbiota in C57BL/6J mice fed with a high-fat diet. Food & Function. 2019;**10**:2847-2860

[155] Seo DB, Jeong HW, Cho D, Lee BJ, Lee JH, Choi JY, et al. Fermented green tea extract alleviates obesity and related complications and alters gut microbiota composition in diet-induced obese mice. Journal of Medicinal Food. 2015;**18**(5):549-556

[156] Carrera-Quintanar L, López Roa RI, Quintero-Fabián S, Sánchez-Sánchez MA, Vizmanos B, Ortuño-Sahagún D. Phytochemicals that influence gut microbiota as prophylactics and for the treatment of obesity and inflammatory diseases. Mediators of Inflammation. 2018:9734845

[157] Zheng J, Zhou Y, Li Y, Xu DP, Li S, Li HB. Spices for prevention and treatment of cancers. Nutrients. 2016;**8**:495

[158] Zhang Z, Chen Y, Xiang L, Wang Z, Xiao GG, Hu J. Effect of curcumin on the diversity of gut microbiota in ovariectomized rats. Nutrients. 2017a;**9**:1146

[159] Kang C, Wang B, Kaliannan K, Wang XL, Lang HD, Hui SC, et al. Gut microbiota mediates the protective effects of dietary capsaicin against chronic lowgrade inflammation and associated obesity induced by high-fat diet. MBio. 2017;**8**:e00470-173

[160] Shen W, Shen M, Zhao X, Zhu H, Yang Y, Lu S, et al. Anti-obesity effect of capsaicin in mice fed with high-fat diet is associated with an increase in population of the gut bacterium Akkermansia muciniphila. Frontiers in Microbiology. 2017b;**8**:1-10

[161] Hamer HM, Jonkers D, Venema K, et al. Review article: The role of butyrate on colonic function. Alimentary Pharmacology & Therapeutics. 2008;**27**(2):104-119

[162] Krautkramer KA, Kreznar JH, Romano KA, et al. Diet–microbiota interactions mediate global epigenetic programming in multiple host tissues. Molecular Cell. 2016;**64**(5):982-992

[163] Al-Lahham SH, Peppelenbosch MP, Roelofsen H, et al. Biological effects of propionic acid in humans; metabolism, potential applications and underlying mechanisms. Biochimica et Biophysica Acta. 2010;**1**:1175-1183

[164] Frost G, Sleeth ML, Sahuri-Arisoylu M, et al. The short-chain fatty acid acetate reduces appetite via a central homeostatic mechanism. Nature Communications. 2014;**5**:3611

[165] den Besten G, van Eunen K, Groen AK, et al. The role of short-chain fatty acids in the interplay between diet, gut microbiota, and host energy metabolism. Journal of Lipid Research. 2013;**54**:2325-2340

[166] DiBaise JK, Zhang H, Crowell MD, et al. Gut microbiota and its possible relationship with obesity. Mayo Clinic Proceedings. 2008;**83**(4):460-469

[167] Flint HJ. Obesity and the gut microbiota. Journal of Clinical Gastroenterology. 2011;**45**(Suppl):S128-S132

[168] Schwiertz A, Taras D, Schäfer K, et al. Microbiota and SCFA in lean and overweight healthy subjects. Obesity. 2010;**18**:190-195

[169] Masotti A. Interplays between gut microbiota and gene expression regulation by miRNAs. Frontiers in Cellular and Infection Microbiology. 2012;**2**:137. DOI: 10.3389/fcimb.2012.00137 eCollection 2012

Chapter 5

The Interaction of Gut Microbiota-brain Axis in Relation to Human Health with the Use of Animal Models

Gaythri Thergarajan and Subha Bhassu

Abstract

The human gastrointestinal tract harbors an extremely complex and dynamic microbial community, including archaea, bacteria, viruses and eukaryota. This gut microbiota usually works with the host to promote health but can sometimes initiate or promote disease. Dysbiosis relationship in gut health indicating the role gut microbiota in promoting the development and progression of brain health. The human gut microbiota is a complex and dynamics microbial community that plays an important role in protecting the host against pathogenic microbes, modulating immunity and regulating metabolic processes. The insights can be elucidated with help of latest omics technology and animal model studies.

Keywords: gut microbiota-brain axis, human health, animal models

1. Introduction

The gut microbiome has been widely accepted to be one of the vital factors causing various disease in human. This area of research has become the niche for many scientists from various fields to explore. Most of the research works are focused on elucidating the influence of gut microbiota in the brain development [1]. Diet plays a crucial role in altering the gut microbiota and some studies focuses on understanding how diet alters gut microbiota and its effect on the development or prevention of metabolic, cardiovascular and brain diseases [2]. Animal studies have always been an important tool in the biomedical research. The interactions of the gut microbiota-brain axis have been studied using various animal models. However, most of the animal research has only able to reveal the fundamental theory such as the diversity of microbial community, the potential microbial pathways and the dysbiosis of the gut microbiota due to diet and drug [3]. The inability of translating many great findings into human system is withdrawal point of animal work [3]. Later, the advancement in the omics-technology has pave the way in the identification of metabolic pathways, microbial species, and metabolites that have strong association with the progression and cure of diseases. Here we have reviewed the interaction of the gut microbiota and brain axis, the pathways, and how the gut wealth affect the human health. Researches on microbiome is translatable into treatments that is able to alter the gut microbiome which could transform the common diseases. This paper also reviewed how the animal research and the application of omics technologies has contribute towards inventions of therapies.

2. The bidirectional communication in microbiota-brain axis

Researches on gut microbiota-brain communication focused on its effects towards digestive functions. However, the current interest in microbiology and neuroscience has given way to understand the psychophysiological consequences of gut-brain or brain-gut as a two-way network [3, 4].

The gut microbiota-brain axis is the term referring to the two-way communication between the gut and the brain [5–8]. More importance is given to elucidate the function of microbes in the gut microbiota-brain axis link as the microbiota can be altered intentionally. The exact mechanism of communication between the gut microbiota and brain has yet to be elucidated, however, the multiple pathways have been identified. The gut microbiota possibly causing an effect on the brain function through the nervous system, endocrine system, immune system, and metabolic system [8].

The bidirectional communication is important in analysing the gut-brain signaling pathways which regulate the host brain and behavior [8]. This bidirectional communication pathway is consisting of the central, enteric, and autonomic nervous systems and the hypothalamic-pituitary-adrenal (HPA) axis. These pathways use the metabolites and the by-products of gut microbes as a communication factor [9]. In recent time, active researches on gut microbiota-brain axis targets the main pathways of the vagus nerve system, the immune system, the neuroendocrine systems, the neurotransmitters and the metabolites [10]. The vagus nerve is responsible in making a physical connection of the gut-brain combo, whereby it allows the brain to sense the gut environment. The vagus nerve extends from the brain to the gut, carrying motor signals and controls the internal digestive, heart and respiratory rate. These motor signals are also transferred to the intestinal cells causing an effect on the gut microbiota [11, 12].

Next, the connection between the gut microbiota and the host immune system is another key research area as studies showing inflammation in neurological and metabolic related disorders [13–16]. The development of low-grade systemic inflammation is associated with impairment in immune response and dysbiotic microbiota. The dysbiosis can regulate on both the innate and adaptive immunity and cause an effect on the gastrointestinal tract and throughout the body. This has been clearly proven in autism spectrum disorders (ASD), epilepsy, Alzheimer's disease, Parkinson's disease and cerebrovascular diseases [16].

Recent findings have showed that the gut microbiota triggers the HPA axis. This pathway controls the neuroendocrine system that modulate stress response, mood and emotions [17]. Evidences shown that microbiota controls the gut hormones and then later regulates the hormone responsible for stress, mood and emotions [17–19]. Gut hormones proven to involve in the physiological processes causing anxiety and depression [18]. A disruption in this bidirectional pathway has been linked with depression, irritable bowel syndrome (IBD) [19] and obesity [18]. These evidences clearly show that gut hormones are potentially regulating the well-being of the host.

The gut microbiota on the other hand, has a major function in the metabolic pathway, which involves energy homeostasis and metabolite production. Animal studies have shown evidence on the ability to produce and metabolise a range of neurotransmitters [1]. A number of neurotransmitters which function as hormones including dopamine, serotonin, noradrenaline, gamma-aminobutyric acid (GABA) has been identified in the context of gut microbiota and brain axis network. These are also known as the hormone-like neurotransmitters which are not only produced in the gut but they do play role in the microbiota. Various factors such as diet, drug, or disease can potentially change the composition of gut microbiota and at instant alters the hormones [20]. In the context of diet, the composition and activity of the gut microbiota can be majorly affected due to the type of food consumed by the host [20].

3. Gut microbiota affecting human health

The alterations of the gut microbiota have the potential to affect the human health and causes various common health as well as major disorders. Firstly, studies have shown the link between the gut microbial community with the common metabolic diseases including obesity, type-2 diabetes, non-alcoholic liver disease, cardio-metabolic disease, and malnutrition [21]. This study has attempted to reveal the connection between abnormal gut microbiota composition and it by products to the dysmetabolism in the diseases mentioned earlier.

The number of cases related to obesity has increased tremendously in the developed countries over the past years [22]. Individuals with obesity has been reported with low microbial gene richness with a relative increase in adiposity, resistance towards insulin, and inflammation [23]. The use of antibiotics before and during pregnancy or in childhood may cause a receding microbial richness of infant and children, increasing the chances of acquiring early-onset of obesity [24, 25]. It was not proven that the receding microbial community is the primary causal factor of obesity, however, it has been shown that low microbial gene richness could be improved with dietary interventions [26].

Type 2 diabetes (T2D) and prediabetes have potential link with altered gut microbiota. An epidemiological study comparing individual without colectomy and patients with total colectomy showed a higher risk of acquiring T2D [27]. This disease has been showing an increased prevalence, especially targeting the adult population and leading to endocrine disorders [28]. Studies have been targeting the products of gut microbiota which may involve in elevating the glucose level in blood. In another study, the gut microbiota of prediabetes individuals shown that there is a reduction in number of *Akkermansia muciniphila* and increase in the number of bacteria pro-inflammatory potentials [29, 30]. *A. muciniphila* is a butyrate-producing bacterium, the reduction on its abundance in the gut may lead to aggravation of opportunistic pathogens [31, 32]. The challenge in revealing the significance of altered gut microbiome to T2D is that the patients are heavily medicated where that would be another main factor causing a dysbiosis to the gut microbiome. That is the reason for using prediabetes individuals as the drug-naïve targets [29].

The gut microbial dysbiosis can also be linked to cardio-metabolic diseases (CMD). Study [33] reported an increase level of *Enterobacteriaceae* and oral cavity species in the gut microbiota of individuals with CMD compared to healthy controls. The microbiota of these individuals has reduced *Bacteriodes* spp. and anti-inflammatory species. In another report, a dysbiotic gut shows potential link to ischaemic heart failure with an elevated level of genes responsible in the systhesis of Lipopolysaccharides [34]. This shows that a disruption in the gut microbiome leads to heightened fatty tissues in the host. A sequencing study done by [35] showed a link between microbiota and atherosclerosis, and later, trimethylamine N-oxide (TMAO), a metabolite from the gut microbiome found to be the causal link to CMD [36].

The microbiota-brain interaction clearly shown is effects on the progression of brain disorders. The development of Parkinson's disease has been linked with formation of protein misfolds in alpha synuclein caused by *Escherichia coli*. *E. coli* was found to produce curli, a protein which causes misfold in other proteins and this error is transmitted to the brain via the vagus nerve [37]. The onset of the ASD has been suggested to cause by segmented filamentous bacteria in the gut. Occurrence of infection during pregnancy causes the bacteria to trigger the T-helper cells to produce immune molecules which later travels to the fetus's brain and provoke autism like behaviors [20].

4. Animal models of gut microbiota research

Intense animal research is intended to gain insight into understanding the reason why there are obvious differences in the human gut microbiota acquired by the healthy and unhealthy individuals. Although the context of the gut microbiota and brain axis is new, it has been well acknowledged in recent time. It is known that gut microbiota regulates the gut metabolism, and various animal studies have revealed that gut microbiota majorly affects the host immune system as well [38]. A number of animal models have been very crucial in enhancing the understanding towards the gut microbiota-brain axis relationship. Yet there are some disadvantages of using the same method.

First of all, the mouse model which would be the common animal model as it can be a good control for age, gender, diet and treatment factors [3, 39–41]. A study uses the *Lactobacillus rhamnosus* to cause region-dependent alterations in the mouse brain, showed neurochemical and behavioral effects [39]. An alteration in the GABA (γ-aminobutyric acid), the main neurotransmitter of the central nervous system was witness, causing an implication on the pathogenesis of anxiety and depression. However, in the vagotomized mice, the effect was not found, indicating the vagus pathway to be the major pathway between the gut and the brain [39]. In this study the vagotomized mouse model being used well to identify the role of bidirectional pathways. Genetic mouse models are also available to target gene specific manipulation. Many studies using mouse model have revealed the influences on the neurophysiology and behavior, cognition, anxiety and depression related issues. However, the translation of research finding using a mouse model on human is difficult. This is very similar to rat model as well. Studies which target the link between gut microbiota and stress uses hamster model [42, 43]. Hamsters on the other hand are difficult to evaluate as they always live in isolation, which allow them to develop metabolic disorders.

Other than mammals, there are also non-mammalian models such as zebrafish [44–47] and *Caenorhabditis elegans* [48, 49]. *C. elegans* has a specialized microbiome is abundance with bacterial taxa where the presence and the number of bacterial taxa found in each individual worm vary from each other. So, the real challenge of working with this organism is to determine the stability and the connection of its microbial community with the host [48]. Many studies revealed the interaction between the microbiota the host can be achieved by using *C. elegans* as a model organism. In another research paper [50], the effects of host environments on bacterial gene expression was successfully studies using the tractable genetic model, *C. elegans*. In this study, the *E. coli* grown *in vitro* were fed to the host, revealed that the host genetics alters the metabolic pathways of the host. The availability of genetic manipulation is the best feature of *C. elegans* model as this could complement the analysis of individual bacterial taxa. A forward and reverse (two-sided) genetic analysis allows the possibility to characterize the microbial processes and its interaction between the host [48].

Zebrafish on the other hand, has been a well establish model animal in the biomedical research, yet the use of this organism in the gut microbiota research has only happened recently. The sequencing method using the bacterial 16s RNA genes revealed the microbial community comprising the bacterial phylum Proteobacteria, Firmicutes and Fusobacteria at all the life cycle stages of zebrafish [44]. Recent studies, have manage to understand the link between the host, microbes and immune response. It has been suggested that gene editing technology may work by targeting a specific gene-deficient in zebrafish to enhance the understanding of immune responses [51]. This animal can be useful in elucidating the conserved molecular mechanisms as they possess similar gene expression and regulation even with different when the organism is isolated from different environment and having varying physiology [51].

Animal	Influence on brain and behavior	Disadvantages
Mouse	Influences the neurophysiology and behavior such as cognition, anxiety and depression	Translation difficulties on human
Rat	Influences the neurophysiology and behavior such as cognition, anxiety and depression	Live in isolation and develop metabolic disorders
C. elegans	Non-mammalian model for validation Tractable genetic model, allow analysis on genetic manipulation	Translation difficulties on human is very high
Zebrafish	Non-mammalian model for validation Reveals the immune response of host Good model for genetic analysis	Translation difficulties on human is very high

This table summarizes animal models used in gut microbiota-brain axis research.

5. Omics in microbiota

The advancement in understanding the interaction between gut microbial community and its host is only possible with recent microbial genomics. The main omics techniques, including metagenomics, metataxonomics, metatranscriptomics, and metabolomics allows the exploration of this area of research. At the initial stages of research, bacterial gene analysis was allow relying on the 16S rRNA sequencing method. Scientists were targeting the conserved region of the nucleotide sequence and compared that with other reference sequences to identify the type of bacterial species in the gut [52, 53]. However, sequencing with 16S provides less information about the functional microbial community in the gut which did not allow the studies to make a correlation between microbes and its potential effects causing failures in experiments [53]. This method was mostly targeting the gut bacterial community but not the other type of microbes such as archea and viruses.

Later, the sequencing method was complemented by the metagenomics approach, where the whole genomic content was accessed using the microbial DNA. Reference genes were used to compare the similarities against the newly available genomic data to identify the functions of the genes coding for the new microbial community [54]. This approach could provide important information on all types of microorganism including archaea, fungi, and viruses at their strain level [55, 56]. However, this approach was not sufficient to understand the functional microbial community at the DNA level, it was needed to translate into functional proteins. Thus, metagenomics was accompanied by metatranscriptomic analysis by translating microbial DNA into RNA [57]. The RNA was later translated into proteins and analysis on microbial functions was continued using metaproteomics. This approach was found to be more comprehensive as it could differentiate between metabolically active microbes in the gut [58]. Mass spectrometry is being used in metaproteomics to measure the expressed proteins which is the important for most biological processes. This information is vital when studying the *in vivo* host-associated microbiomes interactions [57].

In addition, to shed light on the identification of microbial activities in dense, microbial metabolites were targeted by using a tool known as metabolomics. Metabolomic uses techniques such as nuclear magnetic resonance (NMR) spectroscopy or mass spectroscopy (MS) to measure the metabolites present in the gut. Studies has shown that MS is more sensitive in identification of metabolites compared to NMR [59]. The metabolites act as the signaling markers in the communication between the host and its microbiome. As such, imbalance in the intestinal metabolites can be a factor towards development of disease in the host [60]. The various omics technologies explained earlier has been summarized in **Figure 1**.

In the presence of all these omics technologies, scientist believe that they could identify the correlation of microbiome with important human diseases. The gut microbiome influences health, due to the interactions with the immune system.

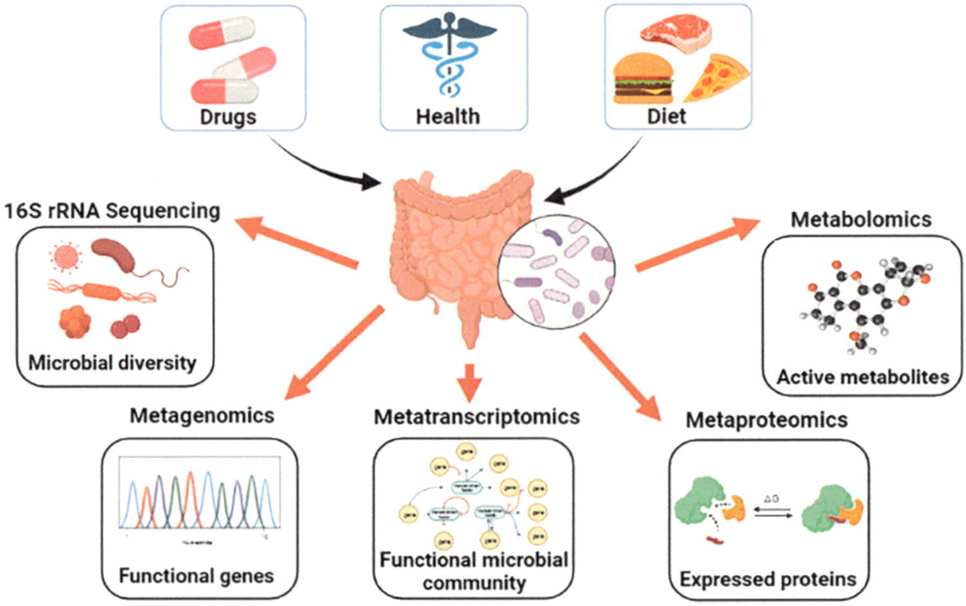

Figure 1.
The use of various omics technologies in revealing the gut microbiota and brain axis relationship.

Understanding the microbial signals will allow new ways to tackle disease. But it is not as easy as it sounds, where more than 50% of the human gut microbiome is yet to be elucidated.

6. Future directions: therapeutic interventions

It is important to identify the key gaps and needs in gut microbiota-brain axis research to plan the future directions and therapeutic interventions. Scientists are only beginning to understand the network between gut microbiota-brain axis. Unraveling the modulation of gut microbiota on the brain health, increases the potential for improving the quality of human life and well-being. The gut microbiome responses to the external factors such as diet and drugs. Drugs are able to modulate the gut microbiome. An integrated understanding on the interaction between the drugs and gut microbiome using the meta-omics technologies can be a major approach towards drug treatment and usage of drugs on certain diseases. There should be rapidly growing studies towards the drug-microbiome interactions targeting available drugs in the markets [61].

Most studies in this field has only attempted using animal models. This method is time consuming, expensive and the findings are difficult to be translated on the human subjects. Culturing the human microbiome by ex vivo culturing together with the meta-omics approach allows development microbiome assays for rapid testing on drug microbiome interactions [57]. Future studies should be focused on understanding the immunological effects of human gut microbes and their role in brain disorders, mapping of neurotransmitters produced by gut microbiota and effect of microbes in early brain development by using human subjects [9]. These interventions are focused in providing nutritional and therapeutic strategies and likely to improve the human quality of life. In most cases when it comes to brain disorders, it is unlikely that these finding provide a permanent cure, however by having the knowledge of these bidirectional communication between the gut microbiota and the brain axis, early predictions or strategies in altering the microbiome to slow down the process would be definitely possible [9].

Nutritional strategies can also be another great practice and are even already on the market, including foods and supplements which help to improve mood, sleep and stress. For instance, altering the diet plan for a child with ASD, could influence the gut microbiota in providing a comfort to gastrointestinal irritation and calm anxiety and hyperactivity. It could be even possible to use probiotics as a complement to drug and therapy for disorders such as schizophrenia. So far many successful trials have achieved by showing the efficacy of probiotics in both strain-specific and disease-specific clinical cases [62]. Studies showed that probiotic supplements are able to benefit the host by producing high bacterial count and the antibiotic therapy could cause a reestablishment of the host microbiome [63, 64].

7. Conclusions

Many advance technologies and animal studies have revealed many interesting facts in elucidating the communication between the gut microbiota and brain axis. However, the fact that most of the studies failed to show the translation of their research finding into human subject is the major gap to be filled in area of research.

Thus, future direction of gut microbiota-brain axis research should focus on the mapping of human gut microbes and they byproducts and finding the immunological effects on the brain disorders. There should be more intervention and preclinical studies focusing on human subjects. The direct link between the human gut microbiota and brain can be only achieved if the bidirectional pathways are revealed from researches focusing on human population.

Acknowledgements

We would like to acknowledge the TRGS grant entitled TR001B-2018A awarded by Ministry of Higher Education. (MOHE) to Prof Suresh as the Program Leader for TRGS and we duly acknowledge to his leadership for this project.

Conflict of interest

The authors declare no conflict of interest.

Author details

Gaythri Thergarajan and Subha Bhassu*
Faculty of Science, Animal Genetics and Genome Evolutionary Biology Lab, Microbiology and Molecular Genetics, Institute of Biological Sciences, University of Malaya, Kuala Lumpur, Malaysia

*Address all correspondence to: subhabhassu@um.edu.my

References

[1] Strandwitz P. Neurotransmitter modulation by the gut microbiota. Brain Research. 2018;**1693**:128-133

[2] Grosso G. Nutritional Psychiatry: How Diet Affects Brain through Gut Microbiota. Nutrients. 2021;**13**(4):1282

[3] Cryan JF, O'Riordan KJ, Cowan CS, Sandhu KV, Bastiaanssen TF, Boehme M, et al. The microbiota-gut-brain axis. Physiological Reviews. 2019;**99**:1877-2013

[4] Agustí A, García-Pardo MP, López-Almela I, Campillo I, Maes M, Romaní-Pérez M, et al. Interplay between the gut-brain axis, obesity and cognitive function. Frontiers in Neuroscience. 2018;**12**:155

[5] Rhee SH, Pothoulakis C, Mayer EA. Principles and clinical implications of the brain–gut–enteric microbiota axis. Nature Reviews Gastroenterology and Hepatology. 2009;**6**(5):306-314

[6] Microorganisms M-A. The impact of the gut microbiota on brain and behavior. Nature Reviews Neuroscience. 2012;**13**(10):701-712

[7] Montiel-Castro AJ, González-Cervantes RM, Bravo-Ruiseco G, Pacheco-López G. The microbiota-gut-brain axis: Neurobehavioral correlates, health and sociality. Frontiers in Integrative Neuroscience. 2013;7:70

[8] Wang H-X, Wang Y-P. Gut microbiota-brain axis. Chinese Medical Journal. 2016;**129**(19):2373-2380

[9] Chakrabarti A, Geurts L, Hoyles L, Iozzo P, Kraneveld AD, La Fata G, et al. The microbiota–gut–brain axis: Pathways to better brain health. Perspectives on what we know, what we need to investigate and how to put knowledge into practice. Cellular and Molecular Life Sciences. 2022;**79**(2):1-15

[10] Sandhu KV, Sherwin E, Schellekens H, Stanton C, Dinan TG, Cryan JF. Feeding the microbiota-gut-brain axis: Diet, microbiome, and neuropsychiatry. Translational Research. 2017;**179**:223-244

[11] Agostoni E, Chinnock J, Daly MDB, Murray J. Functional and histological studies of the vagus nerve and its branches to the heart, lungs and abdominal viscera in the cat. The Journal of Physiology. 1957;**135**(1):182

[12] Prechtl JC, Powley TL. B-afferents: A fundamental division of the nervous system mediating homeostasis? Behavioral and Brain Sciences. 1990;**13**(2):289-300

[13] Coyle PK. Dissecting the immune component of neurologic disorders: A grand challenge for the 21st century. Frontiers in Neurology. 2011;**2**:37

[14] Fung TC. The microbiota-immune axis as a central mediator of gut-brain communication. Neurobiology of Disease. 2020;**136**:104714

[15] Villumsen M, Aznar S, Pakkenberg B, Jess T, Brudek T. Inflammatory bowel disease increases the risk of Parkinson's disease: A Danish nationwide cohort study 1977-2014. Gut. 2019;**68**(1):18-24

[16] Salvo-Romero E, Stokes P, Gareau MG. Microbiota-immune interactions: From gut to brain. LymphoSign Journal. 2020;7(1):1-23

[17] Bao A-M, Swaab DF. The human hypothalamus in mood disorders: The HPA axis in the center. IBRO Reports. 2019;**6**:45-53

[18] Sun L-J, Li J-N, Nie Y-Z. Gut hormones in microbiota-gut-brain cross-talk. Chinese Medical Journal. 2020;**133**(07):826-833

[19] Farzi A, Fröhlich EE, Holzer P. Gut microbiota and the neuroendocrine system. Neurotherapeutics. 2018;**15**(1):5-22

[20] Petra AI, Panagiotidou S, Hatziagelaki E, Stewart JM, Conti P, Theoharides TC. Gut-microbiota-brain axis and its effect on neuropsychiatric disorders with suspected immune dysregulation. Clinical Therapeutics. 2015;**37**(5):984-995

[21] Fan Y, Pedersen O. Gut microbiota in human metabolic health and disease. Nature Reviews Microbiology. 2020;**19**:55-71

[22] World Health Organization. Obesity: Preventing and managing the global epidemic. WHO Technical Series. 2000;**894**

[23] Le Chatelier E, Nielsen T, Qin J, Prifti E, Hildebrand F, Falony G, et al. Richness of human gut microbiome correlates with metabolic markers. Nature. 2013;**500**(7464):541-546

[24] Ajslev T, Andersen C, Gamborg M, Sørensen T, Jess T. Childhood overweight after establishment of the gut microbiota: The role of delivery mode, pre-pregnancy weight and early administration of antibiotics. International Journal of Obesity. 2011;**35**(4):522-529

[25] Mor A, Antonsen S, Kahlert J, Holsteen V, Jørgensen S, Holm-Pedersen J, et al. Prenatal exposure to systemic antibacterials and overweight and obesity in Danish schoolchildren: A prevalence study. International Journal of Obesity. 2015;**39**(10):1450-1455

[26] Cotillard A, Kennedy SP, Kong LC, Prifti E, Pons N, Le Chatelier E, et al. Dietary intervention impact on gut microbial gene richness. Nature. 2013;**500**(7464):585-588

[27] Jensen AB, Sørensen TI, Pedersen O, Jess T, Brunak S, Allin KH. Increase in clinically recorded type 2 diabetes after colectomy. eLife. 2018;**7**:e37420

[28] Association AD. Diagnosis and classification of diabetes mellitus. Diabetes Care. 2014;**37**(Supplement_1):S81-S90

[29] Allin KH, Tremaroli V, Caesar R, Jensen BA, Damgaard MT, Bahl MI, et al. Aberrant intestinal microbiota in individuals with prediabetes. Diabetologia. 2018;**61**(4):810-820

[30] Zhong H, Ren H, Lu Y, Fang C, Hou G, Yang Z, et al. Distinct gut metagenomics and metaproteomics signatures in prediabetics and treatment-naïve type 2 diabetics. EBioMedicine. 2019;**47**:373-383

[31] Qin J, Li Y, Cai Z, Li S, Zhu J, Zhang F, et al. A metagenome-wide association study of gut microbiota in type 2 diabetes. Nature. 2012;**490**(7418):55-60

[32] Karlsson FH, Tremaroli V, Nookaew I, Bergström G, Behre CJ, Fagerberg B, et al. Gut metagenome in European women with normal, impaired and diabetic glucose control. Nature. 2013;**498**(7452):99-103

[33] Jie Z, Xia H, Zhong S-L, Feng Q, Li S, Liang S, et al. The gut microbiome in atherosclerotic cardiovascular disease. Nature Communications. 2017;**8**(1):1-12

[34] Cui X, Ye L, Li J, Jin L, Wang W, Li S, et al. Metagenomic and metabolomic analyses unveil dysbiosis

of gut microbiota in chronic heart failure patients. Scientific Reports. 2018;**8**(1):1-15

[35] Koren O, Spor A, Felin J, Fåk F, Stombaugh J, Tremaroli V, et al. Human oral, gut, and plaque microbiota in patients with atherosclerosis. Proceedings of the National Academy of Sciences. 2011;**108**(Supplement 1):4592-4598

[36] Koeth RA, Wang Z, Levison BS, Buffa JA, Org E, Sheehy BT, et al. Intestinal microbiota metabolism of L-carnitine, a nutrient in red meat, promotes atherosclerosis. Nature Medicine. 2013;**19**(5):576-585

[37] Travagli RA, Browning KN, Camilleri M. Parkinson disease and the gut: New insights into pathogenesis and clinical relevance. Nature Reviews Gastroenterology and Hepatology. 2020;**17**(11):673-685

[38] Zhu B, Wang X, Li L. Human gut microbiome: The second genome of human body. Protein and Cell. 2010;**1**(8):718-725

[39] Bravo JA, Forsythe P, Chew MV, Escaravage E, Savignac HM, Dinan TG, et al. Ingestion of *Lactobacillus* strain regulates emotional behavior and central GABA receptor expression in a mouse via the vagus nerve. Proceedings of the National Academy of Sciences. 2011;**108**(38):16050-16055

[40] Desbonnet L, Clarke G, Shanahan F, Dinan TG, Cryan J. Microbiota is essential for social development in the mouse. Molecular Psychiatry. 2014;**19**(2):146-148

[41] Desbonnet L, Clarke G, Traplin A, O'Sullivan O, Crispie F, Moloney RD, et al. Gut microbiota depletion from early adolescence in mice: Implications for brain and behaviour. Brain, Behavior, and Immunity. 2015;**48**:165-173

[42] El-Ansary A, Bacha AB, Bjørklund G, Al-Orf N, Bhat RS, Moubayed N, et al. Probiotic treatment reduces the autistic-like excitation/inhibition imbalance in juvenile hamsters induced by orally administered propionic acid and clindamycin. Metabolic Brain Disease. 2018;**33**(4):1155-1164

[43] Partrick KA, Chassaing B, Beach LQ, McCann KE, Gewirtz AT, Huhman KL. Acute and repeated exposure to social stress reduces gut microbiota diversity in Syrian hamsters. Behavioural Brain Research. 2018;**345**:39

[44] Roeselers G, Mittge EK, Stephens WZ, Parichy DM, Cavanaugh CM, Guillemin K, et al. Evidence for a core gut microbiota in the zebrafish. The ISME Journal. 2011;**5**(10):1595-1608

[45] Fu J, Gong Z, Bae S. Assessment of the effect of methyl-triclosan and its mixture with triclosan on developing zebrafish (*Danio rerio*) embryos using mass spectrometry-based metabolomics. Journal of Hazardous Materials. 2019;**368**:186-196

[46] Davis DJ, Bryda EC, Gillespie CH, Ericsson AC. Microbial modulation of behavior and stress responses in zebrafish larvae. Behavioural Brain Research. 2016;**311**:219-227

[47] Borrelli L, Aceto S, Agnisola C, De Paolo S, Dipineto L, Stilling RM, et al. Probiotic modulation of the microbiota-gut-brain axis and behaviour in zebrafish. Scientific Reports. 2016;**6**(1):1-9

[48] Dirksen P, Marsh SA, Braker I, Heitland N, Wagner S, Nakad R, et al. The native microbiome of the nematode *Caenorhabditis elegans*: Gateway to a new host-microbiome model. BMC Biology. 2016;**14**(1):1-16

[49] Berg M, Stenuit B, Ho J, Wang A, Parke C, Knight M, et al. Assembly of the *Caenorhabditis elegans* gut microbiota from diverse soil microbial environments. The ISME Journal. 2016;**10**(8):1998-2009

[50] Chan JP, Wright JR, Wong HT, Ardasheva A, Brumbaugh J, McLimans C, et al. Using bacterial transcriptomics to investigate targets of host-bacterial interactions in *Caenorhabditis elegans*. Scientific Reports. 2019;**9**(1):1-12

[51] López Nadal A, Ikeda-Ohtsubo W, Sipkema D, Peggs D, McGurk C, Forlenza M, et al. Feed, microbiota, and gut immunity: Using the zebrafish model to understand fish health. Frontiers in Immunology. 2020;**11**:114

[52] Hillmann B, Al-Ghalith GA, Shields-Cutler RR, Zhu Q, Gohl DM, Beckman KB, et al. Evaluating the information content of shallow shotgun metagenomics. Msystems. 2018;**3**(6):e00069-e00018

[53] Joseph TA, Pe'er I. An introduction to whole-metagenome shotgun sequencing studies. Methods in Molecular Biology. 2021;**2243**:107-122

[54] Forster SC, Kumar N, Anonye BO, Almeida A, Viciani E, Stares MD, et al. A human gut bacterial genome and culture collection for improved metagenomic analyses. Nature Biotechnology. 2019;**37**(2):186-192

[55] Costea PI, Coelho LP, Sunagawa S, Munch R, Huerta-Cepas J, Forslund K, et al. Subspecies in the global human gut microbiome. Molecular Systems Biology. 2017;**13**(12):960

[56] Gevers D, Kugathasan S, Denson LA, Vázquez-Baeza Y, Van Treuren W, Ren B, et al. The treatment-naive microbiome in new-onset Crohn's disease. Cell Host and Microbe. 2014;**15**(3):382-392

[57] Zhang X, Li L, Butcher J, Stintzi A, Figeys D. Advancing functional and translational microbiome research using meta-omics approaches. Microbiome. 2019;**7**(1):154

[58] Heintz-Buschart A, May P, Laczny CC, Lebrun LA, Bellora C, Krishna A, et al. Integrated multi-omics of the human gut microbiome in a case study of familial type 1 diabetes. Nature Microbiology. 2016;**2**(1):1-13

[59] Emwas A-HM. The strengths and weaknesses of NMR spectroscopy and mass spectrometry with particular focus on metabolomics research. In: Metabonomics. New York, NY: Humana Press; 2015. pp. 161-193

[60] Scott TA, Quintaneiro LM, Norvaisas P, Lui PP, Wilson MP, Leung K-Y, et al. Host-microbe co-metabolism dictates cancer drug efficacy in *C. elegans*. Cell. 2017;**169**(3):442-456 e18

[61] Saad R, Rizkallah MR, Aziz RK. Gut pharmacomicrobiomics: The tip of an iceberg of complex interactions between drugs and gut-associated microbes. Gut Pathogens. 2012;**4**(1):1-13

[62] McFarland LV, Evans CT, Goldstein EJ. Strain-specificity and disease-specificity of probiotic efficacy: A systematic review and meta-analysis. Frontiers in Medicine. 2018;**5**:124

[63] Sharp RR, Achkar J-P, Brinich MA, Farrell RM. Helping patients make informed choices about probiotics: A need for research. The American Journal of Gastroenterology. 2009;**104**(4):809

[64] Martín R, Langella P. Emerging health concepts in the probiotics field: Streamlining the definitions. Frontiers in Microbiology. 2019;**10**:1047

Section 3

Infant Microbiota

Chapter 6

Could Alterations in the Infant Gut Microbiota Explain the Development of Noncommunicable Diseases from the DOHaD Perspective?

Maria Esther Mejia-Leon, Alejandra Argüelles-Lopez, Paulina Briseño-Sahagun, Sandra V. Aguayo-Patron and Ana Maria Calderon de la Barca

Abstract

Obesity and its complications are a global public health problem with increasing childhood prevalence. The developmental origins of health and disease (DOHaD) theory explain the maintenance of health or disease development throughout life, related to early life exposures. Although it arises from epidemiological observations, its support for epigenetics is strong. In this chapter, we address the importance of maternal diet in prenatal development, as well as the establishment of the infant microbiota and its postnatal regulating factors. According to the DOHaD theory, breastfeeding and other environmental factors are modulators or enhancers of the epigenetic mechanisms, which explain the increased incidence of noncommunicable diseases. We will discuss the molecular mechanisms related to the microbiota products, their effects on gene expression, and the pathophysiology of the disease. Finally, we will raise the areas of opportunity in childhood for preventive purposes, including the potential role of the use of prebiotics, probiotics, synbiotics, and postbiotics in early life.

Keywords: DOHaD, microbiota, epigenetics, diet, obesity, chronic noncommunicable diseases

1. Introduction

Obesity and chronic noncommunicable diseases, such as diabetes and cardiovascular disease, lead to the main causes of disability and premature mortality worldwide. In recent decades, the prevalence of obesity in the world has increased exponentially in children and adolescents, going from 0.7% to 5.6% in boys and from 0.9% to 7.8% in girls, between 1975 and 2016 [1]. Simultaneously, the incidence of type 2 diabetes (T2D) in the youth increased from 9 to 12.5 cases per 100,000 between 2003 and 2012 [2]. Additionally to the increase in obesity and diabetes, the development of unhealthy habits, such as inadequate diet and sedentary lifestyle in young people, have contributed to the development of cardiovascular diseases (CVD) at an early age [3]. Data from the National Health and Nutrition Examination Survey (NHANES) estimated a prevalence of ischemic heart disease of 0.5–0.6% in the United States for the period 2011–2014 in young adults between 20 and 39 years old. This trend is increasing, and it is expected that by 2030, 43.9% of the US adult population will have some type of cardiovascular disease [4].

The attempts to prevent or palliate the current wave of obesity and the following noncommunicable diseases should be funded at the beginning of human life. An interesting hypothesis is proposed: The Developmental Origins of Health and Disease (DOHaD) that is derived from the Barker hypothesis, which proposed that nutrition during the intrauterine period and exposure to infections after birth determine susceptibility to disease and death from coronary artery disease. This hypothesis has evolved, and currently, critical periods have been identified in fetal life and early childhood, which will determine growth, metabolism, neurogenesis, and future disease risk, expanding the hypothesis to other disorders, such as obesity, diabetes, cardiovascular disease, allergies, and neurological alterations, throughout the life. The DOHaD concept is based on epigenetics and explains the possibility of variations in the programming of the fetus and the infant through the modification of environmental factors, such as diet and infections, in these window periods [5].

Another main component involved in the early life stages is the gut microbiota, defined as the microbial ecosystem that colonizes the gastrointestinal tract, depending on perinatal and environmental factors, such as diet. Its balance is associated with health and its imbalance with the presence of various diseases, although the mechanisms involved have not been fully elucidated; and as with the DOHaD theory, window periods have been identified where its modulation is possible, especially in the perinatal period and up to preschool age [5, 6]. Thus, the aim of this chapter is to discuss the role of the perinatal maternal and infant diet and the gut microbiota to explain the development of chronic noncommunicable diseases from the DOHaD perspective, as key factors in the modulation of epigenetic programming mechanisms, to identify the areas of opportunity for preventive purposes in early childhood.

2. Establishment of the first gut microbiota and its modulating factors

Gut microbiota establishment is determined by several perinatal factors, including gestational pathologies, type of birth, type of feeding, prenatal and perinatal use of antibiotics, complementary feeding, and environmental pollutants [7]. From gestation to the first 2 years of life, these events influence the establishment of the microbiota. Hence, it affects the metabolic and immune response and has a subsequent impact on human health [8].

In the last century, the paradigm dictated that the womb was a sterile environment and that the first microbiota colonized the newborn at the birth time [9]. Even though this is yet a discussion topic, there is evidence pointing toward prenatal exposure to microbes [10]. Despite these, reports of low bacterial abundance and diversity and, in most cases, the lack of culturable bacteria leads to a reasonable doubt about whether it is an established microbiota or only transient exposure to DNA or microbial products that is occurring in the womb [11].

The first major event in microbial colonization for the newborn occurs at birth. Type of birth determines the first gut microbiota composition. Vaginally delivered infant's fecal microbiota is enriched with *Bifidobacterium*, *Bacteroides*, *Clostridium*, and *Lactobacillus* genus. On the other hand, cesarean section is related to a higher abundance of Firmicutes and a lower abundance of Actinobacteria and Bacteroidetes [12]. In the first case, inoculum came mainly from vaginal maternal microbiota, whilst in the second case proceeded from skin and environment, presenting a high abundance of *Staphylococcus*, *Streptococcus*, or *Propionibacteria* [13]. These abundance differences decreased approximately at 6 months of age [12]. Depending on the birth way, different bacterial communities have a competitive advantage, thus first colonizers in infants born by cesarean section delay the establishment of other specific bacterial taxa [14].

First microbiota evolves to adapt to the biochemical environment and in a dependent way on the nutrient availability in the gut [14]. In this sense, whether the infant is breastfed or not, impacts the gut microbiota composition. Ho *et al.* [15], in a meta-analysis study, found higher bacterial diversity and abundance of Bacteroidetes and Firmicutes in non-exclusively breastfed infants compared to those exclusively breastfed at 6 months of age. At the genus level, *Bacteroides*, *Eubacterium*, *Veillonella*, and *Megasphaera* are more abundant in non-exclusively breastfed infants. Bäckhed *et al.* [13] also described differences at genus level between the microbiota of exclusively breastfeed and bottle-feed infants at 4 months of age. The first ones had a microbiota predominated by several species of *Lactobacillus*, whilst the second ones had a high abundance of *Clostridium difficile*, *Granulicatella adiacens*, *Citrobacter* spp., *Enterobacter cloacae*, and *Bilophila wadsworthia*.

Breastfeeding meets all the infant macro and micronutrient requirements during the first 6 months, besides human milk oligosaccharides have a probiotic effect promoting a healthy gut microbiota. Also, human milk provides bioactive compounds that favor immune development, such as immunoglobulins, leukocytes, and antimicrobial peptides. Moreover, human milk harbors its own microbiota, the genera with potential probiotic use as *Lactobacillus, Bifidobacterium*, and *Streptococcus* have been identified as its members [16].

Before 6 months of age, microbial metabolic pathways related to carbohydrate metabolism are higher in non-exclusively breastfed infants [15]. Once other foods aside from breast milk are introduced into the infant diet, functional shifts toward polysaccharides and protein metabolism occur in gut microbiota. However, these changes are not noticeable until breastfeeding cessation. Microbiota composition turns to an adult-like profile with a high abundance of *Bacteroides*, *Bilophila*, *Roseburia*, *Clostridium*, and *Anaerostipes*; but if exclusive breastfeeding continues, *Lactobacillus* and *Bifidobacterium* are higher in gut microbiota at 12 months of age [13].

The evidence on prenatal and perinatal factors influencing the composition of the gut microbiota highlights the importance of microbial colonization as a critical process in early human life. Healthy microbiota is indispensable for immune system shaping and development, and its metabolites promote the integrity of the intestinal mucosa.

3. Maternal diet and its relationship with epigenetics and infant microbiota

Maternal diet is key for offspring development and future disease risk, and this is mediated by epigenetic modifications. In the gestational stage, maternal diet influences offspring epigenetics directly, after birth this influence continues through breastfeeding. Breast milk composition contributes to epigenetics directly as well as through the gut microbiota, which also modulates infant health and development.

During pregnancy, maternal nutrition is a determinant for in utero development, birth weight, and future disease risk. This has been confirmed in studies on the Dutch famine (1944–1945), a period of severe shortage of food in the Netherlands, which have shown that maternal undernutrition during gestation had lasting consequences on the offspring's health. Also, prenatal exposure to the Dutch famine had transgenerational effects highlighting the influence of maternal nutrition over offspring epigenetics [17].

Epigenetic modifications are heritable biochemical markers in the genome that will not change its sequence but will determine gene expression, adapting to diverse environmental factors [18]. There are several epigenetic mechanisms, including DNA methylation, histone modification, and miRNA. DNA methylation is the most studied mechanism, and it relies on one-carbon metabolism. This pathway consists of two cycles, one dependent and one independent of folate. In the first cycle, folate acts as a methyl donor where homocysteine is re-methylated to form methionine. In the second cycle, betaine, and its precursor choline, act as methyl donors. Through this pathway, methionine is turned into S-Adenosylmethionine, the universal methyl donor, which will contribute to DNA methylation [19]. Therefore, DNA methylation depends on methyl donor supply, such as folate, choline, and betaine.

Dietary sources of methyl donors vary according to culture and geographic region. The best sources in Western diets are meat, dairy, and grains; while, in Mediterranean diets, fish, legumes, whole grains, and vegetables are the main sources [20]. According to Taylor *et al.* [21], in Australian preschool children's diet, grains and dairy products were the main sources of folate and choline. Redruello Requejo *et al.* [19] found that the most common sources of one-carbon donors in Spanish pregnant women were animal-source foods, grains, and vegetables. Additionally, culture influences maternal diet in pregnancy and lactation, increasing or decreasing methyl donor intake.

In the gestational stage, offspring's DNA methylation patterns are formed, and maternal intake of methyl donors contributes to proper development and growth. Pauwels *et al.* [22] found that maternal intake of folate, choline, and betaine in the periconceptional stage was associated with methylation of genes related to growth (IGF2), metabolism (RXRA), and appetite (LEP) in 6 months old infants. Insulin-like growth factor II (IGF2) contributes to cell growth and differentiation. According to Xiao *et al.* [23], newborns with fetal growth restriction had a decreased DNA methylation of IGF2. The gene LEP is responsible for leptin production, a hormone that signals appetite regulation and energy expenditure, and LEP methylation is associated with weight gain in the first 10 years of life [24].

There are strong interactions between maternal dietary intake and offspring DNA methylation and health. A high maternal betaine status during pregnancy is associated with lower offspring adiposity; in contrast, a low maternal folate status is associated with a future risk of childhood overweight and obesity [25, 26]. These findings highlight the impact of maternal nutrition during gestation on the offspring's metabolic health.

After birth, the maternal diet continues, influencing DNA methylation through breastfeeding. Therefore, breastfed infants have higher DNA methylation in childhood, compared with formula-fed children [18]. In addition, Briollais *et al.* [27] found that exclusively breastfed infants had more DNA methylation variations, and these were associated with slower BMI growth in the first 6 years of life. The study by Sherwood *et al.* [24] confirmed these findings, where breastfeeding was associated with methylation of LEP and BMI trajectories in childhood. Differently from infant formulas, some breast milk components influence offspring DNA methylation, appetite, and growth; and these components are partly determined by maternal diet.

Compounds such as lipids, oligosaccharides, B vitamins, and betaine, are influenced by dietary intake [28]. Changes in breast milk content will have an impact on infant health, growth, and the development of gut microbiota. Fat and energy content in breast milk is associated with adipose tissue gain in breastfed infants [29]. Additionally, the intake of methyl donors through breast milk could have a direct effect on DNA methylation or could modulate epigenetic modifications via the infant gut microbiota. In two different populations, it was found that betaine concentration in breast milk was associated with infant growth in the first years of life, and betaine concentration was related with the abundance of *Akkermansia municiphila* in the infant's gut, a specie associated with infant growth [30, 31]. Although the evidence is limited, it opens the possibility for the infant gut microbiota to be a modulator of epigenetic modifications.

The development of the gut microbiota occurs in early life, and breast milk has the optimal composition for promoting its proper establishment. For instance, a study that evaluated the fecal microbiota of exclusive breastfed and formula-fed infants found that formula-fed children had a rapid maturation of the gut microbiota, which is associated with future obesity risk [32]. In addition, different types of breastfeeding have an impact on the gut microbiota, breastfed infants with skin-to-skin contact have a healthier microbiota than those fed from a bottle [33].

Many aspects of health are determined by the early gut microbiota, including infant growth. Children with a rapid maturation of the gut microbiota and a high abundance of *Bacteroides* spp. have rapid growth in the first year of life [31]. According to Forbes *et al.* [32], children that were weaned before 6 months old had a rapid maturation of gut microbiota and a greater risk of being overweight at 1 year old. In contrast, the abundance of *Bifidobacterium* and *Akkermansia* at 1-month-old was associated with proper growth in the first year of life [31]. Growth velocity in infancy is a determinant for future metabolic health, and these effects of the gut microbiota on infant growth could be mediated by epigenetic modifications.

The gut microbiota produces a great number of metabolites that participate in epigenetic regulations. Butyrate and propionate produced by *Akkermansia muciniphila* and other species modulate cell transcriptional factors and genes related to lipid metabolism in a murine model [34]. In addition, the effect of Bifidobacterium on infant growth could be mediated by epigenetics, since this genus produces folate, a methyl donor for DNA methylation [35]. Changes in the concentration of the microbiota metabolites could influence post-translational changes in DNA and histones. Therefore, gut microbiota alterations could negatively affect the epigenetic regulation in enterocytes and other cell groups, which in turn will influence infant metabolic health [35].

Epigenetics play a big role in determining infant development and health, and from conception to the postnatal stage, the maternal diet is key for supplying nutrients and components that are necessary for epigenetic regulation. During lactation, breast milk influences epigenetics directly or through the gut microbiota. There is a need for more evidence to elucidate the interactions between breast milk composition, infant gut microbiota, and epigenetic modifications; and to emphasize the importance of maternal diet to ensure proper offspring development, health, and minimize future disease risk.

4. Microbiota products and their local and systemic effects

The main physiological effects observed in the host by gut microbiota could be explained by their metabolite production. There are different products identified and the most studied are short-chain fatty acids (SCFA), where acetate, butyrate, and propionate are the most common and with the most known effects [36]. Other metabolites include trimethylamine-N-oxide (TMAO), obtained from compounds containing choline [37, 38]; secondary bile acids [39]; free anthocyanidins and protocatechuic acid, derived from flavonoid anthocyanins [38], and indolepropionic acid, produced from tryptophan [40]. The ones with beneficial effects on host health are SCFA, anthocyanidins, and indole compounds, and we are going to focus on the first ones.

SCFAs are produced in the bowel lumen by fermentation of dietary fiber [41] by anaerobic bacteria such as *Eubacterium*, *Roseburia*, *Faecalibacterium*, *Coprococcus*, and *Bifidobacterium* [38, 39]. Acetate is predominant, representing 60–75% of the SCFA generated [36] and it is produced via acetyl-CoA and the Wood-Ljungdahl pathway [38]. Propionate can be synthesized by succinate, acrylate, and propanediol pathways, and butyrate by the phosphotransbutyrylase/butyrate kinase, accounting for 25% and 15%, respectively [36, 38].

These molecules exercise their effects by direct or indirect pathways [37]. Direct mechanisms include local or systemic effects, where the microbiota-gut-brain axis is the most studied systemic example [42]; and indirect ways include the effects of these metabolites in other microbes that could modify their function [37].

4.1 Local effects

SCFAs are associated with the maintenance of gut epithelium integrity and protection of the intestinal barrier [36, 37]. Their principal mechanism is as an energy source for enterocytes, but also butyrate and indole derivatives have been associated with aryl hydrocarbon receptor (AhR) ligands, a nuclear receptor whose activation is reported to modulate cell proliferation, immune response, gene expression, and epithelial barrier function [43]. This association with a healthy intestinal epithelium had been explained by the "Warburg effect" or "butyrate paradox." Briefly, fiber-rich diets, associated with an increase in SCFA-producing bacteria, induce normal colonocyte proliferation and apoptosis in neoplastic cells, when metabolism is promoted by glucose [37, 38].

Furthermore, butyrate is important for the maintenance of intestinal barrier integrity because increases the expression of tight junction proteins, such as claudin-1, claudin-7, zonula occludens-1 (ZO-1), and ZO-2 [36, 38, 44]. Also, SCFAs can modulate mucin glycoprotein in the mucus layer [45], induce epithelial cell production of RegIIIγ and β-defensins, antimicrobial peptides [46], and reduce luminal pH [36]. All these functions help to avoid the proliferation of pathogenic bacteria and reduce the translocation of molecules to the systemic circulation.

4.2 Systemic effects

Besides local effects, microbiota metabolites can travel across the intestinal epithelium to systemic circulation or the central nervous system. This can impact different cells via extracellular receptors previously known as G protein-coupled receptors (GPRs) 43, 41, 81, 109A, and 91 [37]. For instance, propionate has a high affinity to

GPR41, now called free fatty acid receptor 3 (Ffar3), which modulates cyclic adenosine monophosphate (cAMP); and to GPR43, now Ffar2, which increases the activity of calcium/protein kinase C (PKC) [36]. Butyrate also has activity on GPR 41 and is the only ligand of GPR109A, now hydrocarboxylic acid receptor 2 (HCA2), which also increases cAMP. Depending on the stimulated cells, effects can be seen in the endocrine, immune, and neurologic systems. For example, activation of the HCA2 receptor in dendritic cells and macrophages is associated with stimulation of T cells into the Treg phenotype [47, 48].

SCFAs also act as inhibitors of histone deacetylases (HDACs). When N-acetyl lysine on DNA histones loses its acetyl group, a more tightly wrapped double chain is formed. HDACs are enzymes that remove this acetyl group, altering DNA transcription by limiting access to transcriptional factors [37]. SCFAs can modify the transcription of a broad range of genes by inhibiting HDACs. Besides, butyrate can act as a ligand of nuclear receptor peroxisome proliferator-activated receptor γ (PPARγ) to modulate the transcription of genes associated with lipolysis and adipogenesis [38, 49]. These different pathways help understand the systemic effects that SCFA can have in several organs, depending on which receptor is activated and the dominant SCFA.

4.3 Role of SCFA in inflammation and immune response

The most beneficial effects of SCFAs are associated with an anti-inflammatory profile. They help to regulate cytokine expression, promoting the production of IL-10, and subsequently, differentiation of Treg cells by the Ffar2 mechanism [36, 37]. Besides, due to their capacity for inhibiting HDACs, SCFAs can impede the activation of nuclear factor-kappa β (NF-κB) [38], a protein complex mainly associated with inflammation. When its RelA/p65 subunit is acetylated, NF-κB can increase gene expression of pro-inflammatory cytokines, such as IL17, IL-1b, IL-6, and IL-12 [50], and enhance transcription of growth factors, adhesion molecules, and immune receptors [36]. Altogether, when the production of pro-inflammatory cytokines is reduced and Treg cells are predominant, the immune response is more regulated, and the risk of inflammatory pathologies is decreased.

SCFAs can suppress the NLRP3 inflammasome and promote an adequate immunologic response by directing T cell differentiation in appropriate phenotypes [36]. For example, reducing systemic inflammation in allergic reactions by modification of T helper type 2 cell numbers [37]. Besides, SCFAs are associated with decreased IL-8 in macrophages and neutrophils, TNF-α in mononuclear cells, and nitric oxide synthase in monocytes [51]. Similarly, butyrate can reduce prostaglandin synthesis by inhibiting COX-2 transcription [50]. All these effects help support the anti-inflammatory profile associated with a fiber-rich diet.

Moreover, SCFAs can influence humoral response. In plasmatic cells, acetate can increase retinoic acid conversion from vitamin A, facilitating response to CD4+ T cell and IgA production [47, 52]. Besides, butyrate and propionate favor antigen affinity inhibiting somatic hypermutation and enhancing class-switch DNA recombination in B cells [53]. SCFAs also influence the proliferation and migration of immune cells, not only as energy sources but through MPAK signal transduction and cascades associated with Ffar2 and Ffar3 receptors [51]. HDACs inhibition activity modulates lymphocyte function, increasing Th1, Th17, and innate lymphoid cells2 (ILC2) and ILC3 [47]. In summary, SCFAs not only allow a more balanced immune response but a more efficient and effective one.

SCFAs have proved to impact immune system development in early life. Exposure to SCFAs during the weaning period is associated with a tolerogenic phenotype and lower risk of inflammatory pathologies later in life, improving CD25+ Treg cells, humoral response, and gut epithelium integrity; confirming microbiota's role in immune system development [54].

4.4 Microbiota-gut-brain axis

Microbiota and their metabolites participate in the bidirectional communication between gut and brain, called the microbiota-gut-brain axis [42]. When SCFAs translocate from intestinal epithelium, they can travel by system circulation, immune system, or enteric-cerebral nervous pathway to provoke changes in distal organs [37, 49].

In the nervous system, butyrate is associated with an increase in cholinergic neurons in the gastrointestinal tract to facilitate motility, propionate with sympathetic activation to greater energy expenditure, and acetate with satiety by hypothalamic stimulation [55]. Similarly, along the gastrointestinal tract, there are enteroendocrine cells (EECs) that sense luminal content and release hormones in the systemic circulation. SCFAs can increase the release of glucagon-like peptide 1 (GLP-1) and peptide YY (PYY), affecting appetite signals and influencing weight control [49]. Therefore, SCFAs can modify autonomic functions and behavior, separately from CNS influence [56].

Another mechanism by which SCFAs alter neurological functions is by direct communication through the vagus nerve and enteric nervous system. SCFAs can alter the expression of GABA receptors [49], production of endothelial nitric oxide, anti-inflammatory and pro-inflammatory components in cerebral microcirculation [55], and increase neurogenesis [56]. Likewise, microbiota's metabolites are associated especially with microglia maturation and function, involving Toll-like receptors (TLRs) [49] and blood–brain barrier integrity [55]. These effects on CNS immune cells explain why SCFAs are associated with less risk of neuroinflammatory disorders.

There are still many mechanisms to be elucidated that could explain all the beneficial effects that microbiota's metabolites in eubiosis could have on host health. However, so far, our diet and early life events are one of the most important interventions to secure a healthy immune and neurologic system, through microbiota modulation.

5. Dysbiosis and early noncommunicable diseases

Throughout life, the structure of the intestinal microbiota can be affected by different factors, such as diet, drugs, the host's immune system, and even the intestinal mucosa itself. Changes in the microbiota can be transient or long-lasting. However, most of the time, alterations in multiple factors are required to generate changes in the microbiota that become harmful to health. This is because the microbiota has resilience, also known as the ability to adapt, to some extent, to changes in the availability of nutrients or environmental conditions [57]. However, when negative conditions are maintained over time, for example, when breastfeeding is not provided or when there is an inadequate dietary pattern or lifestyle in the early years of life, a persistent imbalance of bacterial communities is generated, known as dysbiosis [58].

In addition, some elements have been identified that can amplify or drive changes in the microbiota, making the imbalance more evident and leading directly to dysbiosis. Among them are an increase in the richness of bacteriophages with lytic action in the intestinal environment [59] and the secretion of bacteriocins as a bacterial competition strategy in the intestinal ecosystem. Both situations are enhanced when there is some type of stress [60]. For example, oxidative stress also leads to dysbiosis by promoting the increase of specific bacterial communities and causing the activation of the immune system, as well as the development of subclinical inflammation [57]. This, together with the local and systemic effects of imbalanced SCFAs, described in Section 4 of this chapter, links dysbiosis with the pathophysiological processes of some noncommunicable metabolic diseases, such as obesity, T2D, and CVD [58], as is shown in **Figure 1.**

5.1 Obesity

Different studies have confirmed that there is an imbalance in the intestinal microbiota of obese children when compared to healthy children with normal weight. In general, an increase in the Firmicutes/Bacteroidetes (F/B) ratio has been described in some populations [61]; while in others, no differences have been found at the phylum level [62]. In the systematic review by Indiani *et al.* [63], the results of seven high-quality studies were analyzed and a significant association of Firmicutes with body mass index (BMI) was identified. At the genus and species levels, there is greater consensus regarding the increase in abundance of some Bacteroides species, such as *B. fragilis* [64, 65] and *B. eggerthii* [62]. Other studies have also detected microorganisms such as *Methanobrevibacter smithii, Akkermansia muciniphyla, Desulfovibrionaceae, Bifidobacteriaceae*, and *Enterobacteriaceae* associated with obesity in specific populations, but more studies are needed to increase the evidence of these associations in children [63]. Furthermore, it is generally considered that members of the Bacteroidetes family are the best predictors of the BMI z-score than the phylum analysis [66].

The specific mechanisms by which these associations could explain the early development of obesity from the DOHaD perspective are diverse. In the Canadian Healthy Infant Longitudinal Development (CHILD) birth cohort [67], 935 mother-infant dyads were followed from pregnancy through the first 3 years. Their results explain the intergenerational transmission of overweight and obesity, where having an obese mother and being born by cesarean section increases the risk 5 times for obesity at 1 and 3 years. In this model, the abundance of some specific families of Firmicutes, such as Lachnospiraceae, were sequentially associated with the development of obesity. This association increased in children with obese mothers and was even higher in those born by cesarean section.

Bacteria belonging to the phylum Firmicutes are mostly SCFA producers, such as butyrate and acetate. This supports the findings of Riva *et al.* [66], who found a higher production of SCFA in children with obesity, suggesting a higher fermentative activity. Consequently, when this occurs, energy harvest is increased, which favors a positive energy balance, and contributes to overweight and obesity. Despite this, it depends on the type of SCFA. For example, acetate that is absorbed in the intestine can serve as a substrate for de novo lipogenesis in the liver, which contributes to the accumulation of adipose tissue [68] and compromises the integrity of the intestinal barrier, increasing paracellular permeability and inducing inflammation due to bacterial translocation [6]. In contrast, others SCFA, such as butyrate and propionate, which

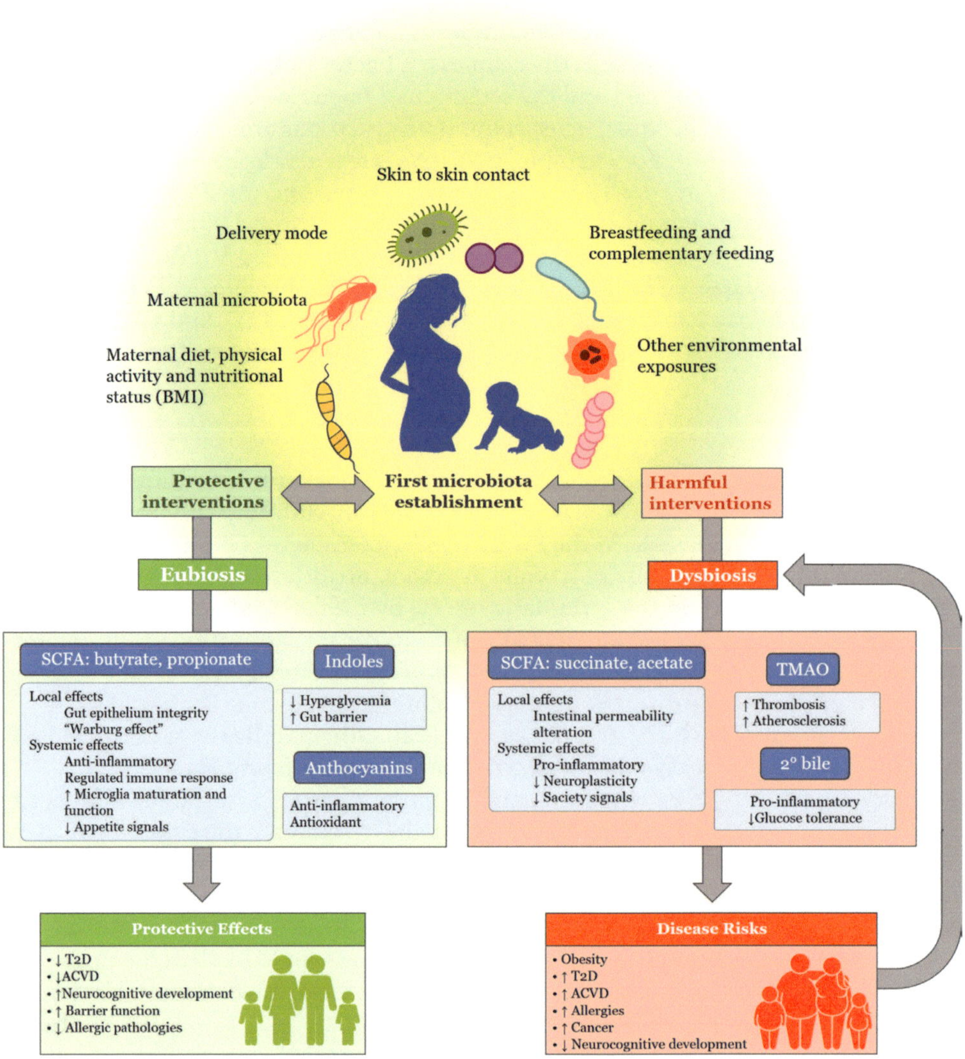

Figure 1.
Perinatal determinants of the first microbiota and effects of protective or harmful interventions for child health through life. BMI: Body mass index, SCFA: Short-chain fatty acids, TMAO: Trimethylamine-n-oxide, T2D: Type 2 diabetes, ACVD: Atheroesclerotic cardiovascular disease.

are dominant products in eubiosis, have a protective effect against obesity. Among the proposed mechanisms, its role in reducing cholesterol synthesis, improving insulin sensitivity, inducing fatty acid oxidation, and leptin gene expression stand out [69].

In obese Canadian children [70], prebiotic supplementation for 16 weeks was associated with a normalized rate of weight gain, decreased percent body fat, and changes in gut microbiota structure, characterized by the increase of *Bifidobacterium* spp. This highlights the role of the microbiota in obesity and the impact that a high-fiber diet could have on its prevention and treatment in childhood.

5.2 Type 2 diabetes

There is increasing evidence of the role of the microbiota in the development of type 2 diabetes (T2D) in youth. In a murine study [71], it was found that during pregnancy, maternal gut microbiota provides protection against obesity and diabetes, through mechanisms related to the SCFA receptors GRP41 and GRP43, which are part of the FFAR family of receptors. This axis participates in the prenatal development of the metabolic and neural systems, driving the development of enteroendocrine cells and pancreatic beta cells. In this way, the deficiency in the signaling of this pathway caused sympathetic dysfunction, compromising energy metabolism, and inducing hyperglycemia.

As in children with obesity, adult patients with T2D have heterogeneous results regarding the F/B ratio [72, 73]. In a study conducted in China [74], it was found that when separating patients with T2D according to the presence or absence of chronic complications, the group without chronic complications presented a higher F/B ratio than those with complications, at the expense of increased Proteobacteria in the latter. Furthermore, some opportunistic pathogens have been identified as part of the microbiota of T2D patients, such as *Bacteroides caccae, Clostridium hathewayi, Clostridium ramosum, Clostridium symbiosum, Eggerthella tarda, and Escherichia coli* [75]. Thus, in general, in patients with T2D, there is a depletion of butyrate-producing bacteria such as *Prevotella* and *Bifidobacterium*. Also, decreased levels of *Akkermansia muciniphila* have been related to mucosal damage and induction of inflammation by activation of the immune system in the lamina propria [74, 75].

Seeking to integrate the previous observations, different mechanisms have been proposed that link the microbiota with the regulation of glycemia. Among them is the production of SCFA due to its effects already described and the increase in the secretion of incretins such as GLP-1 and its role in the differentiation of enteroendocrine cells. In addition, there is evidence regarding their participation in the metabolism of bile acids (BA) and the consequent induction of local and peripheral signals, and the regulation of adipose tissue by promoting white adipose tissue browning and by acting as a trigger for metabolic inflammation [76].

5.3 Cardiovascular disease

The microbiota and its metabolites also modulate the risk and progression of atherosclerosis. Changes in the microbiota diversity and structure have been described in people with atherosclerotic cardiovascular disease (ACVD). As cardiovascular disease is a complication of obesity as well as diabetes, the identified mechanisms coincide with those we have described for these diseases. For example, in a study with 218 patients with ACVD [77], an increased abundance of Enterobacteriaceae and *Streptococcus* spp. was found, with a decrease in butyrate-producing bacteria such as *Prevotella copri* and *Alistepes Shahii*, when compared with the fecal microbiota from 187 healthy controls. Thus, among the associated mechanisms stand out the induction of inflammation, the alteration of lipid metabolism and glucose homeostasis, as well as bacterial translocation. These findings are secondary to alterations in the F/B ratio and in the profile of metabolites such as SCFAs, TMAOs, and BA [78]. In the pathophysiology of ACVD, TMAOs, in particular, have been linked to increased foam cell activation, prothrombotic platelet response, and reverse cholesterol transport, raising the risk of myocardial infarction, stroke, and death [78].

Given that both cardiovascular disease and T2D have a pre-pathogenic period that can last for decades, and because overweight development usually begins in childhood, the perinatal period and early childhood represent a window of opportunity for their prevention and risk modulation.

6. Areas of opportunity in fetal and lactating periods

There are different ways to gauge the window of opportunity during pregnancy and lactation periods. Firstly, the mother's diet and physical activity during the preconceptional and pregnancy periods can induce favorable epigenetic modifications in early life. Second, the delivery way influences the intestinal microbiota composition of the newborn, where the advantage is vaginal delivery, followed by breastfeeding. At this point is quite important the physical contact between mother and child. Exclusive breastfeeding in the first 6 months, depending on the mother's diet, can stimulate the best epigenetic activity to keep a normal growth rate, avoiding a rapid development by direct action or through the intestinal microbiota functionality. After 6 months, a proper food introduction is essential for promoting present and future child's health, and for inducing a favorable intestinal microbiota balance.

As previously described, DNA methylation is crucial for processes epigenetically regulated. In the early embryogenic stage, most parental gametic methylation signs are erased before the acquisition of marks at implantation and beyond. Just after conception, the external environment influences early embryonic events, which are crucial for the DOHaD concept [79]. Therefore, the mother's diet, energy balance, composition as well as her nutritional status, and physical condition are determinant during the periconceptional period [80]. Depending on the nutrient balance and richness of the women's diet in methyl donor compounds, epigenetics modulation will promote normal growth to prevent accelerated fetal growth.

Dietary recommendations during pregnancy are related to amounts of energy, macro-, and micronutrients, such as vitamins and minerals. Dietary reference intake is 340 extra calories during the 2nd pregnancy trimester and 452 for the third one [81]. Pregnant women require a diverse diet, including fruits, vegetables, legumes, nuts, seeds, grains, and tubers, as well as animal-origin products such as dairy, meat, poultry, fish, and eggs. In contrast, pregnant women should avoid some raw seafood, alcohol, and caffeine.

Often health-related practices of a particular cultural group, based on its beliefs, negatively or positively affect the science-based dietary recommendations. For instance, Western diet patterns can fulfill the extra calories for pregnancy, mainly with animal-based products and supplements of vitamins and minerals. Although animal-origin foods contain enough choline, during pregnancy the folate requirement is 600 ug/d and the fiber recommendation is high (28–30 g/d), so supplements are needed [81]. Plant origin fiber is the best recommendation because fruits and vegetables contain in addition to fiber, some very important compounds with antioxidant activity, as well the methyl donor compounds such as betaine and folates, present in leafy green vegetables, broccoli, beans, and peas.

In addition to a diverse and balanced diet, during pregnancy physical activity is necessary for improving glucose tolerance and insulin activity, preventing excessive weight gain. The mother's emotional well-being is important for the fetus, and fitness promotes an easier delivery.

The delivery mode defines the structure of the neonatal microbiota with an advantage of vaginal delivery over C-section delivery, and it is a key factor for the right development of the immune system [82]. The vaginal microbiota is the source of bacterial colonization for the neonate, with implications for the neonate and the mother's health. Before delivery, the vaginal microbiota is mainly dominated by Lactobacillus, and just after delivery, it becomes diverse and similar to the neonatal oral microbiota [83].

In some cases, vaginal delivery is not possible, and C-section is done; additionally, because of strong causes, such as illness and drug treatments, feeding is through milk formulas. Apparently, the window of opportunity is lost, but there are other techniques to ensure healthy microbiota, for example, the use of probiotics, prebiotics, synbiotics, and postbiotics (PPSP) either by the mother or by the newborn. Prebiotics are nondigestible components of food that selectively promote the growth of beneficial bacteria in the intestine; while probiotics are live microorganisms that, administered in adequate amounts, confer a health benefit. On the other hand, synbiotics are a combination of prebiotics and probiotics, while postbiotics are an emerging option, which are soluble products or metabolites (such as SCFA) of commensal bacteria or bacterial components that provide benefits to the host [84]. The use of a combination of strains principally *Bifidobacterium* with or without prebiotics led to an increasing population of bifidobacteria in the newborn microbiota, close to the one of vaginal delivery. Although the effect is larger in breastfeeding children, due to the prebiotic effect of breast milk, even in mixed or formula feeding, there is an additional effect if the intervention begins just after birth [82].

There are still hospitals that pull apart the newborn from the mother if there was a C-section delivery, premature birth, or another reason associated with the mother or newborn's health. Independently of the delivery mode, skin-to-skin contact between mother and child just after the first hour of life improves the possibility of exclusive breastfeeding in the lactation period. This technique helps to reduce neonatal morbidity due to multiple benefits; for instance, stabilizes cardiopulmonary function, and reduces the risk of hypoglycemia, hypothermia, and infections. After delivery, the effect on the mother is a reduction in anxiety and postpartum bleeding [85]. Everywhere, the neonatal intensive care units should promote family participative care, assisting skin-to-skin contact between the mother and newborn as soon as possible, for the establishment of breastfeeding [86].

Regarding general dietary recommendations during breastfeeding, there are higher requirements for carbohydrates and energy intake of up to 500 extra calories, from the beginning to 6 months of lactation. In addition, choline, dietary fiber, and water intake should be higher during the breastfeeding period than during pregnancy [81]. Besides a balanced diet with animal and vegetable sources, mothers should avoid some raw seafood, alcohol, smoking, and caffeine. It is very important to have the best diet for the mother and child's well-being. Installation of breastfeeding is mandatory to induce a good balance of the child's intestinal microbiota, for appropriate immune system development and general child health.

Once and again, microbiota appears in this chapter. It is because the community of different microorganisms in the intestinal tract produces metabolites and cell detritus involved in human metabolic functions. Furthermore, the microbiota influences the immune and central nervous systems; as such, the inhibition of the feeding activity promotes neurons, which ultimately decreases appetite [87]. Therefore, microbiota in dysbiosis could be implicated in metabolic disorders, such as obesity.

A strategy to help the infant to maintain the balance and achieve the stability of its microbiota is to make a correct introduction of solid foods in its diet or complementary feeding. It starts when breast milk or formula composition is not sufficient to accomplish the nutritional requirements of infants, usually from six to 23 months. Breastfeeding can continue at the same time as complementary feeding; the focus is to provide nutrients enough to meet the nutritional requirements of infants. A complementary feeding that is carried out in a staggered manner allows the microbiota to adapt and enrich itself in diversity; thus, it becomes more stable. The problem is that if neglected, complementary feeding has the potential to contribute to childhood overweight and obesity [88].

PPSPs have shown beneficial potential for treating overweight and obesity in children. The proposed mechanism is the modulation of the structure of the microbiota, the profile of microbial metabolites, and the improvement of the intestinal barrier mechanism [84, 89]. In patients with T2D, the use of PPSP decreases fasting blood glucose, total cholesterol, triglycerides, and insulinemia, as demonstrated in the meta-analysis by Bock et al. [90]. However, more studies are still needed to define its role in the prevention and/or treatment of chronic noncommunicable diseases, especially during pregnancy and early childhood.

The task to accomplish a good approach to raising a healthy child across the life course looks so difficult, but any effort pays off with profit. A well-planned pregnancy followed by a balanced and diverse diet, a vaginal delivery with immediate breastfeeding, skin-to-skin contact between mother and newborn, and basic care for the first months is crucial for the metabolic programming of the baby. However, also a carefully complementary feeding from 6 to 24 months, as well as an adequate lifestyle, will help maintain eubiosis, the proper maturation, and functioning of the immune system, and reduce the risk of developing early chronic diseases.

7. Conclusion

There is evidence that demonstrates the relationship between alterations in the intestinal microbiota and the risk of developing chronic noncommunicable diseases throughout life, such as obesity, T2D, and stroke. The involved mechanisms derive from the local and systemic effect of microbiota products, such as SCFAs, indoles, anthocyanins, TMAOs, and BA, as modulators of the inflammatory response and lipid metabolism, among others. Perinatal and early childhood factors modulate the first microbiota and early metabolic programming by epigenetic mechanisms. Thus, the intestinal microbiota is an additional component to the epigenetic mechanisms that strengthen the DOHaD theory and that should be considered in the establishment of preventive measures in the first 1000 days of life.

Conflict of interest

The authors declare no conflict of interest.

Author details

Maria Esther Mejia-Leon[1]*, Alejandra Argüelles-Lopez[2], Paulina Briseño-Sahagun[1], Sandra V. Aguayo-Patron[3] and Ana Maria Calderon de la Barca[2]

1 Mexicali School of Medicine, Autonomous University of Baja California, Mexicali, Baja California, Mexico

2 Nutrition Department, Research Center in Food and Development (CIAD, A.C.), Hermosillo, Sonora, Mexico

3 School of Health Sciences, University of the Valley of Mexico, Hermosillo, Sonora, Mexico

*Address all correspondence to: esther.mejia86@uabc.edu.mx

References

[1] Blüher M. Obesity: Global epidemiology and pathogenesis. Nature Reviews Endocrinology. 2019;**15**(5):288-298. DOI: 10.1038/s41574-019-0176-8

[2] Mayer-Davis EJ, Lawrence JM, Dabelea D, Divers J, Isom S, Dolan L, et al. Incidence trends of type 1 and type 2 diabetes among youths, 2002-2012. New England Journal of Medicine. 2017;**376**(15):1419-1429. DOI: 10.1056/nejmoa1610187

[3] Andersson C, Vasan RS. Epidemiology of cardiovascular disease in young individuals. Nature Reviews Cardiology. 2018;**15**(4):230-240. DOI: 10.1038/nrcardio.2017.154

[4] Benjamin EJ, Blaha MJ, Chiuve SE, Cushman M, Das SR, Deo R, et al. Heart disease and stroke Statistics-2017 update: A report from the American Heart Association. Circulation. 2017;**135**(10):e146-e603. DOI: 10.1161/CIR.0000000000000485

[5] Sarkar A, Yoo JY, Valeria Ozorio Dutra S, Morgan KH, Groer M. The association between early-life gut microbiota and long-term health and diseases. Journal of. Clinical Medicine. 2021;**10**(3):459. DOI: 10.3390/jcm10030459

[6] Mejía-León M, Barca A. Diet, microbiota and immune system in type 1 diabetes development and evolution. Nutrients. 2015;**7**(11):9171-9184. DOI: 10.3390/nu7115461

[7] Butel MJ, Waligora-Dupriet AJ, Wydau-Dematteis S. The developing gut microbiota and its consequences for health. Journal of Developmental Origins of Health and Disease. 2018;**9**(6):590-597. DOI: 10.1017/S2040174418000119

[8] Selma-Royo M, Tarrazó M, García-Mantrana I, Gómez-Gallego C, Salminen S, Collado MC. Shaping microbiota during the first 1000 days of life. In: Guandalini S, Indrio F, editors. Probiotics and Child Gastrointestinal Health: Advances in Microbiology, Infectious Diseases and Public Health. Volume 10. Cham: Springer International Publishing; 2019. p. 3-24

[9] Perez-Muñoz ME, Arrieta M-C, Ramer-Tait AE, Walter J. A critical assessment of the "sterile womb" and "in utero colonization" hypotheses: Implications for research on the pioneer infant microbiome. Microbiome. 2017;**5**(1):48. DOI: 10.1186/s40168-017-0268-4

[10] Coscia A, Bardanzellu F, Caboni E, Fanos V, Peroni DG. When a neonate is born, so is a microbiota. Life. 2021;**11**(2):148. DOI: 10.3390/life11020148

[11] Blaser MJ, Devkota S, McCoy KD, Relman DA, Yassour M, Young VB. Lessons learned from the prenatal microbiome controversy. Microbiome. 2021;**9**(1):8. DOI: 10.1186/s40168-020-00946-2

[12] Rutayisire E, Huang K, Liu Y, Tao F. The mode of delivery affects the diversity and colonization pattern of the gut microbiota during the first year of infants' life: A systematic review. BMC Gastroenterology. 2016;**16**(1):86. DOI: 10.1186/s12876-016-0498-0

[13] Bäckhed F, Roswall J, Peng Y, Feng Q, Jia H, Kovatcheva-Datchary P, et al. Dynamics and stabilization of the human gut microbiome during the first year of life. Cell Host & Microbe.

2015;**17**(5):690-703. DOI: 10.1016/j.chom.2015.04.004

[14] Tamburini S, Shen N, Wu HC, Clemente JC. The microbiome in early life: Implications for health outcomes. Nature Medicine. 2016;**22**(7):713-722. DOI: 10.1038/nm.4142

[15] Ho NT, Li F, Lee-Sarwar KA, Tun HM, Brown BP, Pannaraj PS, et al. Meta-analysis of effects of exclusive breastfeeding on infant gut microbiota across populations. Nature. Communications. 2018;**9**(1):4169. DOI: 10.1038/s41467-018-06473-x

[16] Lyons KE, Ryan CA, Dempsey EM, Ross RP, Stanton C. Breast Milk, a source of beneficial microbes and associated benefits for infant health. Nutrients. 2020;**12**(4):1039. DOI: 10.3390/nu12041039

[17] Veenendaal MV, Painter RC, de Rooij SR, Bossuyt PM, van der Post JA, Gluckman PD, et al. Transgenerational effects of prenatal exposure to the 1944-1945 Dutch famine. BJOG. 2013;**120**(5):548-553. DOI: 10.1111/1471-0528.12136

[18] Mallisetty Y, Mukherjee N, Jiang Y, Chen S, Ewart S, Arshad SH, et al. Epigenome-wide association of infant feeding and changes in DNA methylation from birth to 10 years. Nutrients. 2020;**13**(1):99. DOI: 10.3390/nu13010099

[19] Redruello Requejo M, Carretero Krug A, Samaniego Vaesken ML, Partearroyo Cediel T, Varela MG. Quantification, dietary intake adequacy, and food sources of nutrients involved in the methionine-methylation cycle (choline, betaine, folate, vitamin B6 and vitamin B12) in pregnant women in Spain. Nutrición Hospitalaria. 2021;**38**(5):1026-1033. DOI: 10.20960/nh.03684

[20] NIH. Dietary Supplement Fact Sheets: National Institutes of Health; 2021 [March 23, 2022]. Available from: https://ods.od.nih.gov/factsheets/list-all/

[21] Taylor RM, Smith R, Collins CE, Evans T-J, Hure AJ. Dietary intake and food sources of one-carbon metabolism nutrients in preschool aged children. European Journal of Clinical Nutrition. 2019;**73**(8):1179-1193. DOI: 10.1038/s41430-018-0376-7

[22] Pauwels S, Ghosh M, Duca RC, Bekaert B, Freson K, Huybrechts I, et al. Maternal intake of methyl-group donors affects DNA methylation of metabolic genes in infants. Clinical Epigenetics. 2017;**9**(1):16. DOI: 10.1186/s13148-017-0321-y

[23] Xiao X, Zhao Y, Jin R, Chen J, Wang X, Baccarelli A, et al. Fetal growth restriction and methylation of growth-related genes in the placenta. Epigenomics. 2016;**8**(1):33-42. DOI: 10.2217/epi.15.101

[24] Sherwood WB, Bion V, Lockett GA, Ziyab AH, Soto-Ramírez N, Mukherjee N, et al. Duration of breastfeeding is associated with leptin (LEP) DNA methylation profiles and BMI in 10-year-old children. Clinical Epigenetics. 2019;**11**(1):128. DOI: 10.1186/s13148-019-0727-9

[25] van Lee L, Tint MT, Aris IM, Quah PL, Fortier MV, Lee YS, et al. Prospective associations of maternal betaine status with offspring weight and body composition at birth: The growing up in Singapore towards healthy outcomes (GUSTO) cohort study. The American Journal of Clinical Nutrition. 2016;**104**(5):1327-1333. DOI: 10.3945/ajcn.116.138818

[26] Wang G, Hu FB, Mistry KB, Zhang C, Ren F, Huo Y, et al. Association

between maternal prepregnancy body mass index and plasma folate concentrations with child metabolic health. JAMA Pediatrics. 2016;**170**(8):e160845. DOI: 10.1001/jamapediatrics.2016.0845

[27] Briollais L, Rustand D, Allard C, Wu Y, Xu J, Rajan SG, et al. DNA methylation mediates the association between breastfeeding and early-life growth trajectories. Clinical Epigenetics. 2021;**13**(1):231. DOI: 10.1186/s13148-021-01209-z

[28] Samuel TM, Zhou Q, Giuffrida F, Munblit D, Verhasselt V, Thakkar SK. Nutritional and non-nutritional composition of human milk is modulated by maternal, infant, and methodological factors. Frontiers in Nutrition. 2020;7:576133. DOI: 10.3389/fnut.2020.576133

[29] De Fluiter KS, Kerkhof GF, Van Beijsterveldt IALP, Breij LM, Van De Heijning BJM, Abrahamse-Berkeveld M, et al. Longitudinal human milk macronutrients, body composition and infant appetite during early life. Clinical Nutrition. 2021;**40**(5):3401-3408. DOI: 10.1016/j.clnu.2020.11.024

[30] Ribo S, Sánchez-Infantes D, Martinez-Guino L, García-Mantrana I, Ramon-Krauel M, Tondo M, et al. Increasing breast milk betaine modulates akkermansia abundance in mammalian neonates and improves long-term metabolic health. Science Translational Medicine. 2021;**13**(587):eabb0322. DOI: 10.1126/scitranslmed.abb0322

[31] Alderete TL, Jones RB, Shaffer JP, Holzhausen EA, Patterson WB, Kazemian E, et al. Early life gut microbiota is associated with rapid infant growth in hispanics from Southern California. Gut Microbes. 2021;**13**(1):1961203. DOI: 10.1080/19490976.2021.1961203

[32] Forbes JD, Azad MB, Vehling L, Tun HM, Konya TB, Guttman DS, et al. Association of exposure to formula in the hospital and subsequent infant feeding practices with gut microbiota and risk of overweight in the first year of life. JAMA Pediatrics. 2018;**172**(7):e181161. DOI: 10.1001/jamapediatrics.2018.1161

[33] Fehr K, Moossavi S, Sbihi H, Boutin RCT, Bode L, Robertson B, et al. Breastmilk feeding practices are associated with the co-occurrence of bacteria in mothers' milk and the infant gut: The CHILD cohort study. Cell Host & Microbe. 2020;**28**(2):285-97.e4. DOI: 10.1016/j.chom.2020.06.009

[34] Lukovac S, Belzer C, Pellis L, Keijser BJ, De Vos WM, Montijn RC, et al. Differential modulation by akkermansia muciniphila and faecalibacterium prausnitzii of host peripheral lipid metabolism and histone acetylation in mouse gut organoids. MBio. 2014;**5**(4):e01438-14-e. DOI: 10.1128/mbio.01438-14

[35] Mischke M, Plösch T. More than just a gut instinct–the potential interplay between a baby's nutrition, its gut microbiome, and the epigenome. American Journal of Physiology-Regulatory, Integrative and Comparative Physiology. 2013;**304**(12):R1065-R10R9. DOI: 10.1152/ajpregu.00551.2012

[36] Liu P, Wang Y, Yang G, Zhang Q, Meng L, Xin Y, et al. The role of short-chain fatty acids in intestinal barrier function, inflammation, oxidative stress, and colonic carcinogenesis. Pharmacological Research. 2021;**165**:105420. DOI: 10.1016/j.phrs.2021.105420

[37] McCarville JL, Chen GY, Cuevas VD, Troha K, Ayres JS. Microbiota metabolites in health and disease. Annual Review of Immunology.

2020;**38**(1):147-170. DOI: 10.1146/annurev-immunol-071219-125715

[38] Nogal A, Valdes AM, Menni C. The role of short-chain fatty acids in the interplay between gut microbiota and diet in cardio-metabolic health. Gut Microbes. 2021;**13**(1):1-24. DOI: 10.1080/19490976.2021.1897212

[39] Fiorucci S, Distrutti E. Bile acid-activated receptors, intestinal microbiota, and the treatment of metabolic disorders. Trends in Molecular Medicine. 2015;**21**(11):702-714. DOI: 10.1016/j.molmed.2015.09.001

[40] Dodd D, Spitzer MH, Van Treuren W, Merrill BD, Hryckowian AJ, Higginbottom SK, et al. A gut bacterial pathway metabolizes aromatic amino acids into nine circulating metabolites. Nature. 2017;**551**(7682):648-652. DOI: 10.1038/nature24661

[41] Miller TL, Wolin MJ. Pathways of acetate, propionate, and butyrate formation by the human fecal microbial flora. Applied and Environmental Microbiology. 1996;**62**(5):1589-1592. DOI: 10.1128/aem.62.5.1589-1592.1996

[42] Torres-Fuentes C, Schellekens H, Dinan TG, Cryan JF. The microbiota–gut brain axis in obesity. The Lancet Gastroenterology & Hepatology. 2017;**2**(10):747-756. DOI: 10.1016/s2468-1253(17)30147-4

[43] Marinelli L, Martin-Gallausiaux C, Bourhis J-M, Béguet-Crespel F, Blottière HM, Lapaque N. Identification of the novel role of butyrate as AhR ligand in human intestinal epithelial cells. Scientific Reports. 2019;**9**(1):643. DOI: 10.1038/s41598-018-37019-2

[44] Wang H, Shi P, Zuo L, Dong J, Zhao J, Liu Q, et al. Dietary non-digestible polysaccharides ameliorate intestinal epithelial barrier dysfunction in IL-10 knockout mice. Journal of Crohn's and Colitis. 2016;**10**(9):1076-1086. DOI: 10.1093/ecco-jcc/jjw065

[45] Xia B, Wu W, Zhang L, Wen X, Xie J, Zhang H. Gut microbiota mediates the effects of inulin on enhancing sulfomucin production and mucosal barrier function in a pig model. Food & Function. 2021;**12**(21):10967-10982

[46] Zhao Y, Chen F, Wu W, Sun M, Bilotta AJ, Yao S, et al. GPR43 mediates microbiota metabolite SCFA regulation of antimicrobial peptide expression in intestinal epithelial cells via activation of mTOR and STAT3. Mucosal Immunology. 2018;**11**(3):752-762. DOI: 10.1038/mi.2017.118

[47] Kim CH. Control of lymphocyte functions by gut microbiota-derived short-chain fatty acids. Cellular & Molecular Immunology. 2021;**18**(5):1161-1171. DOI: 10.1038/s41423-020-00625-0

[48] Singh N, Gurav A, Sivaprakasam S, Brady E, Padia R, Shi H, et al. Activation of Gpr109a, receptor for niacin and the commensal metabolite butyrate, suppresses colonic inflammation and carcinogenesis. Immunity. 2014;**40**(1):128-139. DOI: 10.1016/j.immuni.2013.12.007

[49] Song X, Wang L, Liu Y, Zhang X, Weng P, Liu L, et al. The gut microbiota–brain axis: Role of the gut microbial metabolites of dietary food in obesity. Food Research International. 2022;**153**:110971. DOI: 10.1016/j.foodres.2022.110971

[50] Torun A, Enayat S, Sheraj I, Tunçer S, Ülgen DH, Banerjee S. Butyrate mediated regulation of RNA binding proteins in the post-transcriptional regulation of inflammatory gene expression. Cellular Signalling.

2019;**64**:109410. DOI: 10.1016/j.cellsig.2019.109410

[51] Li M, van Esch B, Wagenaar GTM, Garssen J, Folkerts G, Henricks PAJ. Pro- and anti-inflammatory effects of short chain fatty acids on immune and endothelial cells. European Journal of Pharmacology. 2018;**831**:52-59. DOI: 10.1016/j.ejphar.2018.05.003

[52] Goguyer-Deschaumes R, Waeckel L, Killian M, Rochereau N, Paul S. Metabolites and secretory immunoglobulins: Messengers and effectors of the host–microbiota intestinal equilibrium. Trends in Immunology. 2022;**43**(1):63-77. DOI: 10.1016/j.it.2021.11.005

[53] Sanchez HN, Moroney JB, Gan H, Shen T, Im JL, Li T, et al. B cell-intrinsic epigenetic modulation of antibody responses by dietary fiber-derived short-chain fatty acids. Nature Communications. 2020;**11**(1):60. DOI: 10.1038/s41467-019-13603-6

[54] Al Nabhani Z, Dulauroy S, Marques R, Cousu C, Al Bounny S, Déjardin F, et al. A weaning reaction to microbiota is required for resistance to immunopathologies in the adult. Immunity. 2019;**50**(5):1276-88.e5. DOI: 10.1016/j.immuni.2019.02.014

[55] Mirzaei R, Bouzari B, Hosseini-Fard SR, Mazaheri M, Ahmadyousefi Y, Abdi M, et al. Role of microbiota-derived short-chain fatty acids in nervous system disorders. Biomedicine & Pharmacotherapy. 2021;**139**:111661. DOI: 10.1016/j.biopha.2021.111661

[56] Koh A, De Vadder F, Kovatcheva-Datchary P, Bäckhed F. From dietary fiber to host physiology: Short-chain fatty acids as key bacterial metabolites. Cell. 2016;**165**(6):1332-1345. DOI: 10.1016/j.cell.2016.05.041

[57] Weiss GA, Hennet T. Mechanisms and consequences of intestinal dysbiosis. Cellular and Molecular Life Sciences. 2017;**74**(16):2959-2977. DOI: 10.1007/s00018-017-2509-x

[58] Belizário JE, Faintuch J. Microbiome and gut dysbiosis. In: Silvestre R, Torrado E, editors. Metabolic Interaction in Infection. Cham: Springer International Publishing; 2018. pp. 459-476

[59] Norman Jason M, Handley Scott A, Baldridge Megan T, Droit L, Liu Catherine Y, Keller Brian C, et al. Disease-specific alterations in the enteric virome in inflammatory bowel disease. Cell. 2015;**160**(3):447-460. DOI: 10.1016/j.cell.2015.01.002

[60] Ghazaryan L, Tonoyan L, Ashhab AA, Soares MIM, Gillor O. The role of stress in colicin regulation. Archives of Microbiology. 2014;**196**(11):753-764. DOI: 10.1007/s00203-014-1017-8

[61] Bervoets L, Van Hoorenbeeck K, Kortleven I, Van Noten C, Hens N, Vael C, et al. Differences in gut microbiota composition between obese and lean children: A cross-sectional study. Gut Pathogens. 2013;**5**(1):10. DOI: 10.1186/1757-4749-5-10

[62] López-Contreras BE, Morán-Ramos S, Villarruel-Vázquez R, Macías-Kauffer L, Villamil-Ramírez H, León-Mimila P, et al. Composition of gut microbiota in obese and normal-weight Mexican school-age children and its association with metabolic traits. Pediatric Obesity. 2018;**13**(6):381-388. DOI: 10.1111/ijpo.12262

[63] Indiani C, Rizzardi KF, Castelo PM, Ferraz LFC, Darrieux M, Parisotto TM. Childhood obesity and firmicutes/bacteroidetes ratio in the gut microbiota:

A systematic review. Childhood Obesity. 2018;**14**(8):501-509. DOI: 10.1089/chi.2018.0040

[64] Ignacio A, Fernandes MR, Rodrigues VAA, Groppo FC, Cardoso AL, Avila-Campos MJ, et al. Correlation between body mass index and faecal microbiota from children. Clinical Microbiology and Infection. 2016;**22**(3):258.e1-258.e8. DOI: 10.1016/j.cmi.2015.10.031

[65] Scheepers LEJM, Penders J, Mbakwa CA, Thijs C, Mommers M, Arts ICW. The intestinal microbiota composition and weight development in children: The KOALA birth cohort study. International Journal of Obesity. 2015;**39**(1):16-25. DOI: 10.1038/ijo.2014.178

[66] Riva A, Borgo F, Lassandro C, Verduci E, Morace G, Borghi E, et al. Pediatric obesity is associated with an altered gut microbiota and discordant shifts in Firmicutes populations. Environmental Microbiology. 2017;**19**(1):95-105. DOI: 10.1111/1462-2920.13463

[67] Tun HM, Bridgman SL, Chari R, Field CJ, Guttman DS, Becker AB, et al. Roles of birth mode and infant gut microbiota in intergenerational transmission of overweight and obesity from mother to offspring. JAMA Pediatrics. 2018;**172**(4):368. DOI: 10.1001/jamapediatrics.2017.5535

[68] Gomes AC, Hoffmann C, Mota JF. The human gut microbiota: Metabolism and perspective in obesity. Gut Microbes. 2018;**9**(4):308-325. DOI: 10.1080/19490976.2018.1465157

[69] Amabebe E, Robert FO, Agbalalah T, Orubu ESF. Microbial dysbiosis-induced obesity: Role of gut microbiota in homoeostasis of energy metabolism. British Journal of Nutrition. 2020;**123**(10):1127-1137. DOI: 10.1017/s0007114520000380

[70] Nicolucci AC, Hume MP, Martínez I, Mayengbam S, Walter J, Reimer RA. Prebiotics reduce body fat and Alter intestinal microbiota in children who are overweight or with obesity. Gastroenterology. 2017;**153**(3):711-722. DOI: 10.1053/j.gastro.2017.05.055

[71] Kimura I, Miyamoto J, Ohue-Kitano R, Watanabe K, Yamada T, Onuki M, et al. Maternal gut microbiota in pregnancy influences offspring metabolic phenotype in mice. Science. 2020;**367**(6481):eaaw8429. DOI: 10.1126/science.aaw8429

[72] Larsen N, Vogensen FK, Van Den Berg FWJ, Nielsen DS, Andreasen AS, Pedersen BK, et al. Gut microbiota in human adults with type 2 diabetes differs from non-diabetic adults. PLoS One. 2010;**5**(2):e9085. DOI: 10.1371/journal.pone.0009085

[73] Sedighi M, Razavi S, Navab-Moghadam F, Khamseh ME, Alaei-Shahmiri F, Mehrtash A, et al. Comparison of gut microbiota in adult patients with type 2 diabetes and healthy individuals. Microbial Pathogenesis. 2017;**111**:362-369. DOI: 10.1016/j.micpath.2017.08.038

[74] Zhao L, Lou H, Peng Y, Chen S, Zhang Y, Li X. Comprehensive relationships between gut microbiome and faecal metabolome in individuals with type 2 diabetes and its complications. Endocrine. 2019;**66**(3):526-537. DOI: 10.1007/s12020-019-02103-8

[75] Cunningham AL, Stephens JW, Harris DA. Gut microbiota influence in type 2 diabetes mellitus (T2DM). Gut Pathogens. 2021;**13**(1):50. DOI: 10.1186/s13099-021-00446-0

[76] Gérard C, Vidal H. Impact of gut microbiota on host glycemic control. Frontiers in Endocrinology. 2019;**10**:29. DOI: 10.3389/fendo.2019.00029

[77] Jie Z, Xia H, Zhong S-L, Feng Q, Li S, Liang S, et al. The gut microbiome in atherosclerotic cardiovascular disease. Nature. Communications. 2017;**8**(1):845. DOI: 10.1038/s41467-017-00900-1

[78] Ahmad AF, Dwivedi G, O'Gara F, Caparros-Martin J, Ward NC. The gut microbiome and cardiovascular disease: Current knowledge and clinical potential. American Journal of Physiology-Heart and Circulatory Physiology. 2019;**317**(5):H923-HH38. DOI: 10.1152/ajpheart.00376.2019

[79] Fleming TP, Sun C, Denisenko O, Caetano L, Aljahdali A, Gould JM, et al. Environmental exposures around conception: Developmental pathways leading to lifetime disease risk. International Journal of Environmental Research and Public Health. 2021;**18**(17):9380. DOI: 10.3390/ijerph18179380

[80] Silver MJ, Saffari A, Kessler NJ, Chandak GR, Fall CH, Issarapu P, et al. Environmentally sensitive hotspots in the methylome of the early human embryo. eLife. 2022;**11**:e72031. DOI: 10.7554/elife.72031

[81] Kominiarek MA, Rajan P. Nutrition recommendations in pregnancy and lactation. Medical Clinics of North America. 2016;**100**(6):1199-1215. DOI: 10.1016/j.mcna.2016.06.004

[82] Martín-Peláez S, Cano-Ibáñez N, Pinto-Gallardo M, Amezcua-Prieto C. The impact of probiotics, prebiotics, and synbiotics during pregnancy or lactation on the intestinal microbiota of children born by cesarean section: A systematic review. Nutrients. 2022;**14**(2):341. DOI: 10.3390/nu14020341

[83] Li H, Jiang J, Nie C, Xiao B, Li Q, Yu J. Community structure and ecological network's changes of vaginal microbiome in women right after delivery. Frontiers in Pediatrics. 2022;**10**:750860. DOI: 10.3389/fped.2022.750860

[84] Li HY, Zhou DD, Gan RY, et al. Effects and mechanisms of probiotics, prebiotics, Synbiotics, and Postbiotics on metabolic diseases targeting gut microbiota: A narrative review. Nutrients. 2021;**13**(9):3211. DOI: 10.3390/nu13093211

[85] Santos APDS, Lamy ZC, Koser ME, Gomes CMRDP, Costa BM, Gonçalves LLM. SKIN-to-skin contact and breastfeeding at childbirth: women's desires, expectations, and experiences. Revista Paulista de Pediatria. 2022;**40**:e2020140. DOI: 10.1590/1984-0462/2022/40/2020140

[86] Lee C-K, Huang X-Y. Psychological processes of postpartum mothers with newborns admitted to the intensive care unit. Asian Nursing Research. 2022;**16**(1):9-17. DOI: 10.1016/j.anr.2021.12.007

[87] Adamantidis A. How the gut talks to the brain. Science. 2022;**376**(6590):248-249. DOI: 10.1126/science.abo7933

[88] Tsenoli M, Khan MAB, Östlundh L, Arora T, Omar O. Complementary feeding practices and the associated risk of childhood obesity among ethnic minority groups living in high-income countries: Protocol for a systematic review and meta-analysis. BMJ Open. 2022;**12**(3):e053821. DOI: 10.1136/bmjopen-2021-053821

[89] Barczynska R, Kapusniak J, Litwin M, Slizewska K, Szalecki M. Dextrins from maize starch as substances activating the growth of bacteroidetes and actinobacteria simultaneously

inhibiting the growth of firmicutes, responsible for the occurrence of obesity. Plant Foods for Human Nutrition. 2016;**71**(2):190-196. DOI: 10.1007/s11130-016-0542-9

[90] Bock PM, Telo GH, Ramalho R, et al. The effect of probiotics, prebiotics or synbiotics on metabolic outcomes in individuals with diabetes: A systematic review and meta-analysis. Diabetologia. 2021;**64**(1):26-41. DOI: 10.1007/s00125-020-05295-1

Section 4

Probiotics and Dairy Products for Microbiota Enhancement

Chapter 7

Probiotics in Processed Dairy Products and Their Role in Gut Microbiota Health

Kishwer Fatima Sherwani and Dil Ara Abbas Bukhari

Abstract

Probiotics are the beneficial microorganisms, catalase negative which restore microbial balance inside the gut of humans as well as animals. *Lactobacillus* the earliest probiotic that have the beneficial impact on health. These "Good Microorganisms" can be obtained not only from various non-dairy products but also from processed dairy products like. Another economically viable method is microencapsulation for preserving probiotics and the stability is improved by glucose. Even the vitamins manufacturer the probiotic bacterial agents. The health benefits of probiotics include increased immunological responses, relief of lactose intolerance symptoms, therapy for diarrhea, reduction in serum of cholesterol, production of vitamin, anticarcinogenic. Probiotics play a wide range in the host body (e.g., decreasing illnesses and stress, enhancing immunity, modulation of gut microbiota, nutritional assistance, improving quality of water, etc.). So, the positive effects of probiotics help to boost animal feed value and growth and improve aquaculture breeding and hatching rates. Probiotics can lower the prevalence and severity of illnesses, showing their promise to cure or prevent COVID-19. *Lactobacillus casei* also interact with epithelial cells with Toll-like receptors (TLRs) to improve the production of cytokines that are important in the enhancement of cell productivity and prevent apoptosis during restoration, which promote survival and proliferation. The preservation of the human GI or lung microbiota might help prevent COVID-19, as dysbiosis plays an essential role in people's vulnerability to infectious illnesses. Most of the experimental studies proved that bacteria isolated from processed dairy products belonged to lactic acid bacteria and are declared as probiotic bacteria. In present review, various research studies regarding significance of probiotics as well as their extraction from processed dairy products are discussed.

Keywords: probiotics, processed dairy products, gastrointestinal diseases, commercial forms, *Lactobacillus*, health benefits

1. Introduction

"Probiotic" is a Greek term which meaning "for life" [1]. Over the years, there have been numerous meanings of the term "probiotic." Probiotic means living microbiota cultures that enhance the qualities of the indigenous microbiota when supplied to people or animals. These bacteria are extensively spread in nature and are suitable for usage in food industries. Some foods are considered a good source of probiotics, including milk products, e.g., yogurt. Milk and milk products are often linked with microorganisms that replenish the digestive tract with helpful maintenance [2]. Different generations of lactic acid bacteria like *Enterococcus species, Lactococcus species*, different strains of *Lactococcus* and *Streptococcus* are also present in processed products of milk.

Lactic acid bacteria are a type of gram-positive bacteria that cannot produces spores and catalase negative, so they characterized in the absence of cytochrome system. Different sources of food like yogurt, milk, cheese that are dairy in nature are the good source of probiotics [2]. Most processed milk products contain the *Lactobacillus,* and other strains of it.

Increased taste of food and better shelf life are among the most evident benefits of LAB fermentation. The fact that LABs can produce the bacteriocins and provide health benefits, including controlling intestinal infections, improving lactose uses, reducing the ammonia level of the blood, and providing effective resistance to gastric acid and bile, makes it generally considered safe for bacteria to be used. Impacts the immunological system and decreases serum cholesterol levels. Probiotics are live, health-effecting micro-organisms when eaten. Different LAB strains, particularly *Lactobacilli* and *Bifidobacterium* which reside in the adult bacteria in the gut of humans with good therapeutic functions have been used increasingly as probiotics.

1.1 History

In 1908, Metchnikoff gave the first probiotic definition, which suggested that use of fermented dairy produce extended the life of the product. In 1956 Lilly and Stillwell decided that some growth stimulators for another microbe were secreted by a microbe. The usage of the word probiotic may lead to this beneficial impact on such micro-organisms.

Parker [1] coined the word probiotic to define the chemicals and organisms that produce microbial balance in the gastrointestinal system. This meaning has given rise to significance meaning, including antibiotics, of the word substances. Fuller refined the description of Parker and described probiotics as a living micro-organism, which has a beneficial effect on warm-blooded animal health through the restoration of native gut microbiota. The definition by Fuller highlighted the survivability and beneficial effects of the probiotic on animals. In the instance of probiotics host, Haveenar and Veld described them as 'viable micro–organics' in 1992, which are administered to humans or animals as mono- or mixed cultures which have a favorable effect on the health of host, by enhancing the intestinal characteristics of the micro-flora. A probiotic is a live bacterium injected in milk products based on the Salminen criteria and enhance host health and diet.

Schaafsma extends the concept and according to its definition probiotics are live bacteria which, if swallowed and absorbed in a more than intrinsic basic nutritional way, have health benefits on the host. Salminen noted that the probiotic in milk products is a microbial culture. Accordingly, the food matrix is a big indicator for the microbe and food being regarded as probiotics. However, because non-dairy food includes viable probiotic products, it was not justifiable to take milk products only as a probiotic matrix. In 2001, Probiotics were seen by Schrezenmeir and by De Vrese as live micro-organisms which have independent health effects of the place of activity.

In 2001, probiotics are known as "live micro-organisms by the WHO (WORLD HEALTH ORGANNIZATION) and FAO (FOOD AND AGRICULTURE ORGANIXATION OF THE UNITED STATES) that provide the host with a health advantage by being given adequately." Finally, in 2014, the ISAPP changed the latter definition of probiotics to somewhat and described it as 'live micro-organisms, which provide a health benefit on a host when administering in appropriate quantities.' This concept has been extensively adopted by the scientific community since then and is a criterion in most government agencies to evaluate medicines, food, and supplements as probiotics. The viability of a microbe and of the related product as 'probiotic' is one of the essential criteria in this definition.

1.2 Lactobacilli

Lactobacillus is regarded as the earliest probiotic reported. This genus is made up of LAB group gram-positive bacteria. These bacteria in rod shape include over 183 recognized species and are frequently used in diverse commercial food processes [3]. *L. acidophilus* is a bacterium that has a positive impact on the host. Recent investigations have investigated the finding of normal vaginal flora of certain species producing hydrogen peroxide. The beneficial products have thus been examined in urine and vaginal tract infections among women, so it helps in the treatment of it. It was also used to treat Candida infections in the mouth (**Figure 1**).

Reticence of harmful organisms like *Salmonella, Shigella* and different *Helicobacter* comprise valuable effects mediated by Lactobacilli. Lactobacilli was also related with

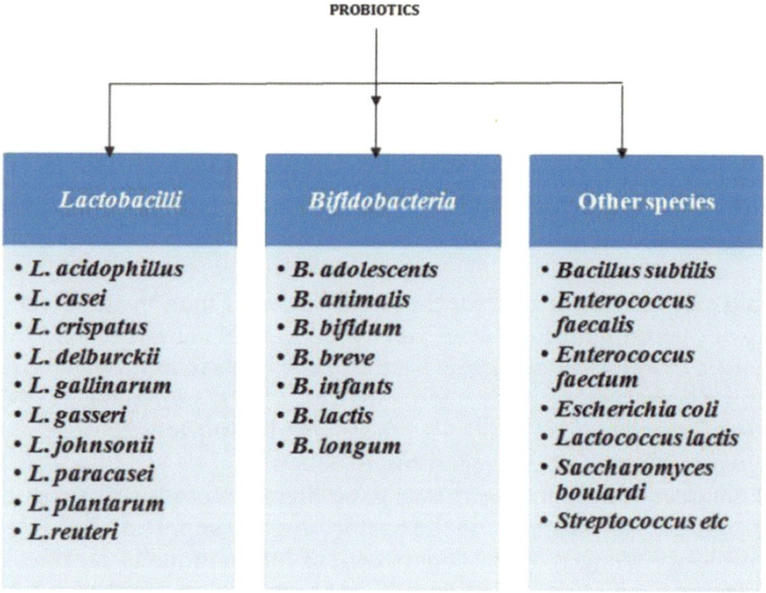

Figure 1.
Types of probiotic strain.

several additional health advantages, e.g., improved immunological response and lessening of lactose intolerance. A good function was also revealed in colon cancer for Lactobacilli. Lactobacilli strengthen the immune system and cure cancer, pancreas, fever blisters, hives. In infants who are born preterm, necrotizing enters colitis (NEC). Cheese is a milk product that may be delivered in the intestine of people by probiotic bacteria. Italian, Argentinean, and Bulgarian cheese have isolated the strains of *L. plantarum*.

1.3 Source of probiotics

Cheese is a milk product that may be delivered in the intestine of people by probiotic bacteria. Italian, Argentinean, and Bulgarian cheese have isolated the strains of *L. plantarum* [4].

1.3.1 Lactic acid bacteria in ice cream

It is referred as a frozen dairy product that is produced by freezing a sterilized mixture to incorporate air to maintain homogeneity. Milk products, sugar, dextrose, water, eggs or egg products, and different non-harmful aromas are used to form a mix that is utilized to make ice cream. Its nutritional content and their energy value are generally determined by their food value [5]. Irrespective of the mix, ice creams usually constitute great sources of dietary energy and are an appropriate substratum for probiotic proliferation.

Ice cream is considered as favorable for the probiotics as it helps in survivability and metabolic activity of probiotics and its different strains shows more benefit for transferring probiotic organisms to the body [6]. Ice creams are more advantageous than other fermented dairy products. But freezing and thawing have detrimental consequences on probiotics, including disruption of metabolic activity and cell death. There have been reports of studies designed to reduce these negative consequences.

Lactobacillus delbrueckii sub specie was investigated by Leandro [7]. Bulgaricus UFV H2b20 is kept at −16°C for 40 days in three formulations of ice cream (it contains less fat, free of fat and high-fat). Although in the three formulations ($P > 0.05$), LABs may be integrated into ice-cream formation and different processing parameters for different consumer groups may be handled (**Table 1**).

1.3.2 Agitated milk products

It is an old method that is used for the preservation of food by the development and activity of microorganisms. Moreover, the additional benefit of fermentation is the formation of metabolites such as bacteriocins, helps in improved nutritional status and for the sensory properties of foodstuffs, and helps in reduced toxic and anti-nutritional components [13]. Due to these positive impacts, these food stuffs are important to human diets since ancient times.

In 2001 regarded probiotics are known to be "living microorganisms by the WHO and FAO that offer health benefits to the host if properly supplied". In 2014, the definition of probiotics changed by the department of International Scientific Association for Probiotics and Prebiotics (ISAPP) as a "lived micro-organism that, when supplied adequately, conveys a health advantage to the host." In 2014, Yerlikaya described that, *Lactobacillus* and varieties of *Bifidobacterium* are the most frequent probiotics in fermented food items. In 2012, Mishra find out that dairy-based matrix is appropriate

Type of ice cream	Name of the probiotics added	Method of probiotic supplementation	Viability and storage conditions (CFU/g)	References
Standard ice cream	*Lactobacillus casei*	The probiotic strain was added after homogenization and heating of ice cream mix	Count from $3.9 \times 10^9$to 3.8×10^8 log CFU/ml 80 days of storage at −20°C	[9]
Standard ice cream	*L. plantarum*	The probiotic strain (encapsulated within a calcium—alginate/chitosan microcapsules containing insulin) was added after homogenization	2.3×10^7 after storage of 90 days at −20°C	[10]
Yog ice cream	*Lactobacillus acidophilus La5*	Encapsulated probiotic strain (alginate-based) was added into the ice cream mix	After 60 days of storage at −18°C, the viable probiotic count was approximately 1×10^7	[11]
Standard ice cream	*L .acidophilus* ATCC 4356	Ice cream mix was fermented with *L. acidophilus* prior to freezing	After 90 days of storage at −18°C the viable probiotic count was 1×10^6	[12]

Table 1.
Use of different strains of probiotics in ice cream [8].

to proliferate probiotics by providing a high carbon and necessary amino acid source owing to the hydrolyzing of lactose and the usage of casein in the proteolytic system.

Traditional white cheese is widely referred to as Lighvan cheese in Tabriz marketplaces. It was originally prepared of raw ewe's milk, raw goats' milk, and raw cow's milk and/or mixed with them from time to time. This sort of cheese is popular and often consumed across Iran and has significant economic and nutritional benefits because of its attractive organoleptic qualities. In the manufacturing and maturation of cheese, lactic acid bacteria (LAB) are involved. Many genera are involved in LAB, such as *Lactobacillus, Streptococcus, Enterococcus and Leucon Stoc.*

1.4 Mechanism of action of probiotic

In 2012, Bermudez studied that the principal mechanisms of probiotics action include to enhance the epithelial barriers, increases in the adherence to gut mucosa and microbial adhesion, generation of antimicrobial compounds and regulation of immune systems. This process is shown in **Figure 2**: that illustrates how these processes appear in the intestinal mucosa, as a schematic illustration. Lactic acid from various carbohydrates (e.g., carbon sources are produced in the micro-organisms of the LAB Group [15]. Pereira [16] conducted a study in which various antibacterial processes of probiotic activity are connected to these components.

1.5 Development of probiotics

As we know that during human development microbiota changes in the gut of human. The newborns' gut is completely sterile yet colonization of several types of

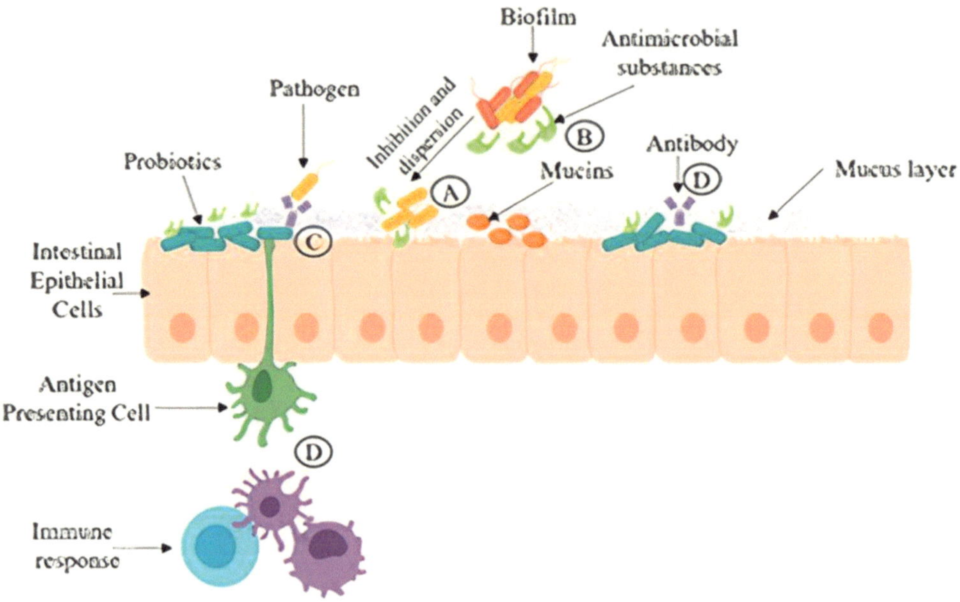

Figure 2.
Probiotics as an alternative antimicrobial therapy: current reality and future directions [14].

bacteria starts shortly after delivery. The first- and second days following delivery have been demonstrated to be present in newborn feces, including coliforms, enterococci, clostridia, and lactobacilli. Bifidobacterial start colonization in three to 4 days and prevail about the fifth day. Coliform numbers drop at the same time. In feces, 1 log count of bifidobacterial is more prevalent in infants than those fed by bottle. Bottle-fed babies show greater levels of strains of Enterobacteriaceae, and other putrefactive bacteria, indicating that babies that are given breast are resistant to gastrointestinal diseases than infants fed by bottles. To ensure a person's diet and health, the gastrointestinal system also modify in addition to the changes in the microbiome that happen throughout human aging. Using antibiotics, for example, might disrupt the balance of gut microbiota, reduce bifidobacterial and lactobacilli count and increase clostridium. This imbalance may result in diarrhea in senior citizens and in those who are immunocompromised.

1.6 Beneficial host response

Some probiotic methods elicit many positive reactions from the host. Most of the effects of these products include: (1) exclusion and competition for pathogen-cell adhesion to epithelia, (2) inborn immune stimulation, (3) compete for nutrients and prebiotically products, (4) manufacture of antimicrobial substances and consequent pathogenic antagonism; DC: dendritic cells; LI: interleukin; M: intestinal cells M. IEC: intestinal cells M (**Figure 3**) [17].

1.7 The probiotics of next generation

The idea of traditional probiotic products, taken from a limited number of microorganisms, is connected to the observation of the health benefits to both humans and

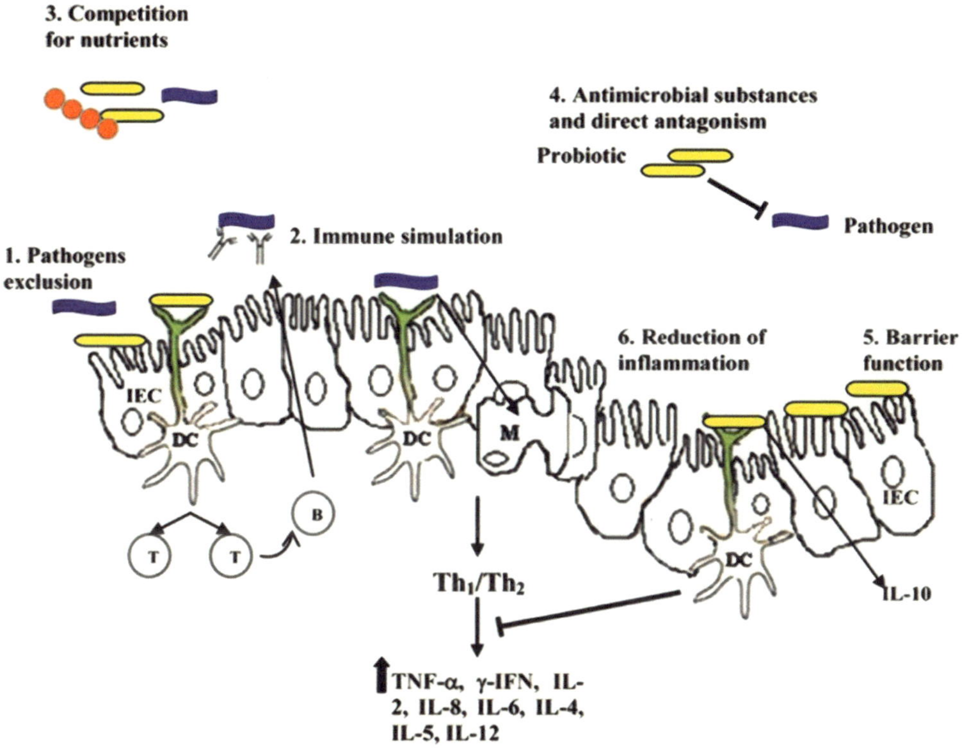

Figure 3.
Probiotics latest advances [17].

animals from the daily consumption of LAB-fermented food. The word 'probiotic' was therefore associated with bacteria that promote health [18]. The development of knowledge of human microbiota of intestine and its importance for disease and health has led to the discovery of several new bacteria that plays a significant role in the health of human through therapeutic modulation of the intestinal microbiota and are called NGP. The study is very much in the interest of investigating the probiotic potential of commensal bacteria. NGPs are defined as "living micro-organisms identified by comparative analyses which provides benefits to health of the host when properly administered".

1.8 Probiotics in fermented milk

The largest existence of probiotic products in the dairy sector is fermented milk. Several studies have successfully applied probiotic strains to milk fermentation and have induced desirable textural properties, that is apart from inducing health-promoting effects. Highly nutritional value makes its widespread availability and the most widely utilized probiotic milk products. Many commercially manufactured probiotics fermented dairy products are commonly used throughout the world (**Table 2**). Gao in 2019 described that the probiotic products of Kefir and Koumiss are the natural fermented milks mostly used in many parts of the world. Many studies have indicated that probiotic strains are incorporated in traditionally fermented milks that aids to improve their positive health impacts. For example, in a natural milk product (lait curd) of Senegal, Parker et al. [19]

Dairy products			
Pediococcus acidilactici SMVDUDB2	Kalarei, a fermented cheese product	80% survival rate at pH 2.0 and 3.0 and 0.3% bile salt concentration, high hydrophobicity affinity (33%) with ethyl acetate, auto aggregation (77.6%), antibacterial activity against *Bacillus subtilis*, *Mycobacterium smegmatis*, *Staphylococcus aureus*, *Proteus vulgaris*, and *Escherichia coli*, and EPS production (2 g/l)	[11]

Table 2.
Probiotics in the dairy industry.

integrated *L. rhamnosus* GG. B. lactic Bi-07 and *L. acidophilus* NCFM, Wang et al. [20] have been integrated into natural milk that aids to improve health of intestine and the immunity of host cell. Many probiotic bacteria in conjunction with traditional probiotic bacteria that are reported to be use in fermented milk to improve the flavor and other characteristics [21, 22], that was linked with functional food. In yogurt, *Streptococcus Thermophilus* and *Lactobacillus delbrücckii* sub specie *bulgaricus, Lactobacillus plantarum* P-8 fermentation have the capacity to improve the yogurt flavor profile by producing 3-methylbutan, acetone, onanal, 2-heptanone, hexanale, (E)-2-octenal and 2-nonanone, compared to controls [23]. In the same way, a high acetic acid, acetoin, 2-butanone, caproic acid, butyric acid, and 2-pentanone content were found in fermented milk containing *L. casei* DN-114001 compared to control group [24].

1.9 Probiotics in yogurt

Yogurt is a functional ingredient that contain probiotics, so there is great interest in producing probiotic yogurts that are either fermented or incorporated into yogurts with different strains of probiotic. Several commercially produced probiotic yogurts are widely utilized worldwide. A standard yogurt is a fermented milk product traditionally made of *L. delbrueckii* subsp. *bulgaricus* and *S. thermophilus* by fermenting the milk. As yogurt starter cultures can survive in human GI tract [25]. These may be considered probiotic because they have health-promoting effects [26]. However, all the strains of the yogurt starter culture worldwide are not identical and, therefore, the probiotic potential of yogurt starter culture in general remains controversial.

The physiochemical, sensory, and microbial characteristics of the yogurts produced by many probiotic strains comparable to the traditionally produced yogurts are much better in many cases. Of course, certain probiotics, such as *L. plantarum* and *L. acidophilus*, can reduce the bisphenol A (estrogenic substance) content of yogurts considerably [27]. The yogurt metabolism leads to large amounts of unmetabolized lactose and residual galactose in yogurts, which have been shown to be more metabolized (full use of lactose and efficient galactose degradation) by probiotic *L. plantarum* WCFS11 [28].

1.10 Probiotic butter and cream

There have been several products in which probiotic products have been incorporated due to its widespread benefits, and butter is also used. This is not limited to fermented milk, yogurt, and cheese. Butter, mainly made up of fats, has many health advantages. Emergent evidence, however, suggests that many cardiovascular diseases and diabetes have a high content of saturated fatty acids in butter [29]. Some probiotic bacteria have been reported to reduce the cholesterol content of cream and butter [30] *(L. casei* subsp. For example, in cultivated cream creams high contents of capric, butyric and caproic acid were produced when a blend of probiotic strains, including *Bifidobacterium bifidum, L. acidophilus, S. thermophilus,* and *L. bulgaricus*, was used in the fermentation of creams enriched by 2% (each) sunflower oil, hazelnut oil and soy oil [31]. The increase of the contents of linoleic and α-linolenic acid in probiotic cream in relation to control cream was observed in another study following Bifidobacterium lactis fermentation [32]. In nondairy butters, for instance the peanut and sunflower-produced butters, cocoa and flaxseed oils, probiotics are more commonly used.

1.11 Probiotics in powdered milk and infant formulas milk powder

Probiotics aid in development of an effective immune system by changing the microflora of the intestine in infants. Probiotics and prebiotics are increasingly added to infant formulae. Probiotic dispensation of bifidobacterial and the strains of lactobacilli in neonatology has developed in worldwide. *B. bifidum* and *L. acidophilus,* dispersed into infant formula (109 CFU/250 mg tablet), have been reported to continue to be more resistant by comparing with breast milk, after storage capacity at 4°C or 6 h [33].

1.12 Technological challenges for dairy products viability

The survival of probiotics is highly crucial as it provide the highly recommended efficiency of probiotics products. During food manufacturing, storage, and gastrointestinal movement, probiotics face multiple stress situations [34]. A minimum of 106 CFU/g of B. bifidum and 107 CFU/g of L. acidophilus in fermented milk are required for several international standards. The probiotic fermented milk should contain at least 107 CFU/ml of live bifidobacteria at the time of consumption in Japan (according to its Association of fermented milk and lactic acid drinking) [35]. The incidence of oxygen in processed dairy products has an impact on most of the probiotic strains and their survival. The oxygen-induced toxicity in milk products poses a major technological obstacle to the development of probiotic fermented milk and yogurt. *Bifidobacterium* species have an anaerobic metabolism of an intestinal origin, which implies that they depend completely on fermentation.

1.12.1 Role of oxygen in viability of yogurt probiotics

The potential reduction of yogurt bifidobacteria after storage in different regions of the world has been documented as oxygen toxicity [36, 37]. Expenditure of probiotics to dissolved oxygen leads to a build-up intracellular of harmful oxygenic metabolites, like superoxide anion [38]. A high level of oxygen is unavoidably included in the product by several processing processes involved in the processing of milk products (e.g., agitation and mixing procedures). In addition, some packing materials during the storage period enable the transmission of ambient oxygen into the food, such as high impact polystyrene packaging—a commonly used packaging material for yogurt worldwide because of its vision, strength, and hygiene levels [39].

The use of oxygen-impermeable containers, process of two-stage fermentation, acid and bile salt resistant strains, probiotics microencapsulation and prebiotics are significant preconditions for enhancing stability and viability of yogurt-based probiotics [40].

1.12.2 Studies demonstrating the viability of probiotics by oxygen

In several investigations, oxygen scavengers or antioxidants have been found to be beneficial agents for increased probiotic viability. Ascorbic acid (Vitamin C), for example, has been observed to improve *L. acidophilus*' survivability in yogurt [41]. Glucose based oxygen scavenger was evaluated for its impact on probiotic development and survival in dairy products [42]. In yogurt prior to fermentation, glucose oxidase (62.25 ppm) was increased to 69.02–86.03% and the number of Bifidobacterium longum by up to 40.32% compared to the control level [43].

1.12.3 Role of glucose for improving probiotic stability

Some probiotic bacteria have a significant pH susceptibility below 3. It is stated that glucose has been included into the growing medium for improving probiotic stability. Corcoran and others observed glucose improving *L. rhamnosus* GG's survival at a pH lower than 3. The addition of GL (1 to 19.4 mM) improved *L. rhamnosus* GG survival in artificial gastric juice from 6.4 to 8 log10 CFU/ml [44]. Therapeutic adaptation of the acid stress was also described as ways to improve Bifidobacterium species stress tolerance and biological characteristics [45].

1.12.4 Probiotic microencapsulation from different manufacturing processes

Another economically viable method is microencapsulation for preserving the probiotics against various treatment procedures and for ensuring their distribution to the human body in a necessary amount. Different studies have indicated improved probiotic survival when embedded. In the case of oligosaccharides, gelatin, inulin and xanthan gum, alginate materials (i.e., sodium alginate and human-like collagen, gelatin-based microspheres, alginate-like gum, gum-Arabic derived from cellulose, maltodextrin, vegetable protein, pectin hydrogel beads, carrageenan, and other proteins) are also supplied with the use of probiotics intended for use in milk products [46].

1.13 Vitamins and probiotics

Vitamins are generally classed to include vitamins (A, D, E and K), fat-soluble or to be water-soluble that including vitamins C, biotin (vitamin H or B7), vitamins B—thiamin (B1) and B—thiamin (B2) and riboflavin (B3) (B12). While fat-soluble vitamins are key cell membrane components, water-soluble vitamins are used as coenzymes, usually conveying chemical groups. People are unable to synthesize most vitamins and must thus be extracted exogenously. Using vitamins can be an alternative to reinforcement using chemically synthetized pseudo-vitamins that is more natural and consumer friendly.

Probiotic bacteria empower a beneficial effect on the host immune system and on the gut microbiota composition and function. In addition, vitamin synthesis has brought various health benefits to the host. Probiotic bacteria, mostly of the *Lactobacillus* and

Bifidobacterium genus, provide several health advantages. The vitamin K, and most aquatic-soluble B vitamins, including as biotin, cobalamins, folates, nicotinic acid, pyridoxine, riboflavin, and thiamine, can manufacture probiotic bacterial agents, members of the gut microbiota, in humans. Probiotic bacteria have been widely investigated to produce B-vitamins, notably folate and riboflavin (B2). Several LAB species manufacture these vitamins, frequently in high quantities, and are therefore often present in fermented foods (e.g., *Lactococcus lactis, Lactobocillus gasseri, and Lactobacillus reuteri)* and *Bifidobacterium* (e.g., *Bifidobacterium adolescentis*). In addition, higher production of vitamins has been achieved through metabolism. Folate biosynthetic genes and biosynthesis operon of riboflavin have been over-expressed in L. lactis, leading to kinds of folate or riboflavin that produce at greater rates. The modified biosynthetic routes of folate and riboflavin in *L. lactis* are used to produce both vitamins simultaneously by directed mutagenesis and selection and metabolic engineering.

1.14 Commercial forms of probiotics

It is possible to absorb probiotic organisms in two primary ways: through fermented meals and through supplements. Fermented foods may come from both dairy and vegetable sources, with yogurt and sauerkraut being the most well-known of each. Freeze-dried (lyophilized) bacteria in powder, pill, or tablet form make up probiotic supplements. For clinical effectiveness, products containing probiotic organisms must contain enough live organisms to exhibit therapeutic benefits, regardless of the way they are ingested. Both fermented foods and supplements can accomplish this feat in the same way and have pros and cons (**Table 3**).

The probiotic strain that has been demonstrated to have the necessary therapeutic effect is essential to achieving successful and repeatable clinical outcomes. *L. rhamnosus* GG, for example, has been proven to prevent viral gastroenteritis and maintain

Delivery system	Pros	Cons
Fermented dairy	• Affordability and easy availability • Ease of incorporation into daily patterns • Additional nutritional benefits • Enhanced bacterial survival	• Contains dairy proteins and lactose • Taste can be issue • Not suitable when traveling • Not suitable for vegans
Capsules	• Ease of administration • Contain no binders	• Not therapeutic in upper GI tract • Many contain allergenic excipients • Higher cost
Tablets	• Ease of administration • Effective in the upper GI tract	• Many contain allergenic or otherwise problematic binders and excipients (e.g., gluten) • Higher cost
Powders	• Effective in upper GI tract • Dosage can be easily adjusted • Can be incorporated into foods or drinks • Contain no binders	

Table 3.
Pros and cons of commercial forms of probiotics [47].

ulcerative colitis in remission, according to research. We cannot assume that other strains of *L. rhamnosus* would behave in a similar fashion. In the same way, a doctor who uses the identical strain used in clinical trials should expect similar outcomes. An effect may be obtained by using a nearly similar strain.

For meals and supplements containing probiotics, the dose depends only on the quantity of live organisms present in the product, not on its composition. In clinical studies, between 107 and 1011 live bacteria per day were used. When administered in a dairy medium, it appears that 100 times less viable bacteria are required to reach the same number of live bacteria in the lower colon. In the upper GI tract, dairy appears to be a good transport medium for the bacteria, boosting their survival.

2. Nutritional requirements, health benefits and probiotics

The health advantages of probiotic dietary items have increased human appeal. Such meal is not only packed in nutrients, but also lowers the risk of many disorders [48]. "Live microbial probiotics, delivered at a sufficient quantity (106 to107 CFU/g), impart health advantages on the host," according to FAO. The medicinal advantages of these products thereby encourage probiotic use [49]. The management of gastrointestinal and urinary tract infections was related with probiotics. Other advantages include enhancing serum cholesterol lactose tolerance levels, increasing host immunity, and preventing antibiotic diarrhea and allergy disorders related with colon cancer [50]. The growing demand for novel probiotic products nevertheless prompted the development of probiotics, these products include ice cream and baby milk powder that supply probiotics.

2.1 Health benefits

The health benefits of probiotics include increased immunological responses, relief of lactose intolerance symptoms, therapy for diarrhea, reduction in serum of cholesterol, production of vitamin, anticarcinogenic. The most important segments of world commerce have been probiotic goods in recent years which can increases annually, the growth rate ranges in 6.8% between the year 2013 and 2018, and then reached to 37.9 billion US$ in 2018. In the meantime, probiotic dairy products are one of the most advanced and a key part of the functional food business [17].

2.2 Probiotic potential of cheese

Cheese is useful because of the high pH, greater amount of fat and from its thick consistency. Cheese is a useful food for probiotics to the gastric intestinal system. The probiotic potential is studied and used as assistant cultures in various kinds of food products or therapeutic preparations a diverse number of food grade lactic acid bacteria (LAB) isolated from PDO cheese from Italian form of Castelmagno product of cheese, Italian and Argentinean cheese products [51]. Lactobacilli are isolated from milk products, in particular from cheese, and have shown a long history of safe use because these micro-organisms are widely used to develop new fermented products, milk or meat, alcoholic beverages and sourdough.

Piewngam [52], studied that the formulations of probiotics are believed to enhance human health, such as immunostimulant effects or interbacterial competition between helpful bacteria and harmful ones. The use of probiotics was seen as a potential method for the prevention and management of many infectious illnesses.

Piewngam et al [53] found a reverse linkage between human colonization with species of Bacillus and *S. aureus*. The researchers also detected a key mechanism through the suppression of quorum sensing through which species of Bacillus can kill *S. aureus*. Chung [54] described that the fengycins are the types of bacilli produced lipopeptides that are identified through the process of chromatography and mass spectrometry.

2.2.1 Efficacy of S. aureus

Another work carried out by Moraes et al. [55] has demonstrated the efficacy of *S. aureus* biofilms produced on titanium discs in *Lactobacillus brevis* and *B. bifidum*. The results indicated a decrease in growth of *S. aureus* on titanium disc when both probiotics were applied but in *L. brevis* strains the highest inhibitory effect was seen.

2.3 Examination of *Helicobacter pylori* (*H. pylori*) as an infection

Goderska et al. [56] examined *Helicobacter pylori (H. pylori)* as an infection which has been seen as difficult to treat, particularly as it has gained an increased resistance to antibiotically widely used. Probiotics in conjunction with antibiotic regimens are increasingly being used to eliminate *H. pylori*. In addition to the advantages of probiotic bacteria to the intestine's probiotics have shown effective for the treatment of various bowel disorders including diarrhea; several positive effects on the stomach, including anti-*Helicobacter pylori* have also been documented [57].

The advantages of probiotic treatment are reduced microbial charge and host tolerance in instances with *H. pylori*. Several research have revealed the positive benefits of various *H. pylori* probiotics by reinforcing the mucosal barrier, while increasing adhesion and immunomodulation competition.

A study carried out by Lahtinen et al. [58] demonstrated that the growth of *Staphylococcus aureus*, often seen in systemic and peri-implant infections, has been prevented from 3 out of 38 strains of *Bifidobacterium*. The antibacterial activity of various probiotic Lactobacilli strains was studied by Lazarenko et al. [59]. In a model of intravaginal infection in mouse, *B. bifidum (B. bifidum)* was found to be largely effective against *S. aureus* with a substantial reduction in the amount of *S. aureus* cells caused by vaginal spraying. In comparison with other probiotic strains of various genera, *B. Bifidum* exhibited superior anti-staphylococcal efficacy.

2.4 Test model of *C. albicans*-infecting mouse to determine the effects of *L. casei* in vaginal candidiasis

In the test model of *C. albicans*-infecting mouse, Liao et al. [60] analyzed the effects of *L. casei* administration in vaginal candidiasis. The animals were inoculated with *L. casei* vaginally throughout 7 days for prophylactic testing. Three mice were killed, and the amount in CFU/ml was measured. The animals had *C. albicans* infected the vaginal cavity 2 days after the infection. The animals were treated with *C. albicans* in therapeutic tests and after 2 days, *L. casesi* was infected for 5 days. The CFU/ml number was then measured in vaginal samples. The findings suggest that prophylactic *L. casei* treatment might enhance vaginal mucosal immunity, increasing IL-17 production during infection. IL-23 levels had also weaker anti-inflammatory effects than those in the control group. In the therapy group, after 5 days of treatment, *L. casei* decreased the fungal vaginal load.

2.5 Probiotics and its advantageous effects on skin

Mottin and Suyenaga [61] described that poor skin problems might impact the quality of life of the patient due to discomfort. Human skin is made up of several fungus and symbiotic bacteria Chronic skin diseases that require lengthy treatment durations and maintenance are acne and atopic dermatitis (AD). In these situations, studies have found satisfactory outcomes without side effects using probiotics. In vitro trials indicate the potential to directly suppress acnes development by producing antibacterial proteins (bacteria) and immunomodulatory effects of probiotics, such as *Streptococcus salivarius* and *Enterococcus faecalis*. It has been demonstrated that probiotics have direct (inhibited *P. acnes*) and indirect (reduce the inflammatory response) advantages [62, 63].

2.6 Influence of probiotics on mental health and disease

Dinan and Cryan [64] think that the intricate bidirectional connection that happens between the brain and gut microbiota (GM) might be a novel approach to determine mental disease treatments. Several studies have found that the GM plays a substantial influence in an individual's mood and behavior, and that it might be very useful in mental health therapy. Stress-induced physiological consequences in the stomach, such as nausea and spells of diarrhea, have a significant impact on the GM balance [65].

Psychobiotics are a novel type of probiotic that is intended to help people with psychiatric illnesses by enhancing their cognitive abilities [66] Many different gut microbial species generate a variety of mood-regulating neuromolecules, which has an impact on host physiology. GABA is produced by *Bifidobacterium* and *Lactobacillus* species, whereas serotonin is produced by *Enterococcus, Escherichia, Streptococcus, and Candida species*, and dopamine is produced by *Bacillus* species [67].

2.7 LAB with in vitro, in situ cholesterol-lowering characteristics

Cholesterol reduction is one of the most favorable properties for probiotic bacteria with lactic acid. In this work, a capability evaluation was carried out of 58 possibly probiotic bacteria containing cholesterol and bile acids for in vitro digestion and cholesterol reductions. The best-performing strains reduced cholesterol levels in broth by 42–55% and were tested in the production of cheese.

In all cheeses, the cholesterol content declined during maturation. The most significant decreases (up to 23%) were obtained by adding *LB. paracasei, paracasei* VC2161 and *Epilithonimonas lactis* BT 161 during cheese-making, all strains were present in the cheese at levels greater than 107 cfu/g up to 60 days after ripening. There was no detrimental influence on the sensory properties of cheese in the adjacent cultures. These strains with demonstrated in vitro characteristics are, therefore, ideal candidates for new probiotic formulations, and can also be utilized to make foods like dairy fermented products effective.

2.8 Probiotic strains

Probiotic microbe selection is based on safety, function, and technology, as described in the following reports. Some probiotic microorganisms are already on the market and have been thoroughly investigated. They must first be able to be produced under industrial circumstances before probiotic strains may be provided to customers. Then, throughout the storage of the crops frozen or freeze-dried as well as food items into which they are formulated, they must survive and keep their functioning. Furthermore, they need to be incorporated into plates without producing flavors or texture. For functional dietary requirements, the following aspects in relation with the probiotic should be considered: Preparing for large-scale manufacturing should be feasible, remain stable and viable for storage and use.

2.8.1 Lactobacillus rhamnosus *GG*

Studies have demonstrated the promotion of immunoregulatory activities by raising regulatory cytokines of interleukin (IL)-10 [68] and the induction of beta (TGF-β) transforming T-cells [69]. In fact, atopic children have proven that the gut microbiota differs from atopic ones. LGG showed a beneficial impact on atopic illness prevention while randomized clinical studies (RCTs) reveal no outcomes [70, 71].

LGG in babies with rotavirus-related diarrhea led to higher increased production of non-specific antibodies and anti-rotavirus antibodies. The neonatal evidence of necrotizing enterocolitis, improving food tolerance and prevention of pathogens colonizing intestine because of competition exclusion, preventing adhesion, and improving mucosal immunoglobulin A (IgA), has shown LGG to be effective in reducing incidences of necrotizing enterocolitis. A new retrospective 6-year cohort study on LGG in extremely small birth weight babies showed the microbiological safety of the strain [72].

2.8.2 Bb12 Bifidobacterium

Bb12 has been available on the market for over 25 years and is one of the best probiotic strains accessible for the most extensive research. It was administered to babies alone or with several different probiotic strains and has demonstrated its well-tolerated and beneficial effects [73]. The Bb12, combined with Streptococcus thermophilus Th4, has already proven that infants are well accepted in a formulation and have decreased the colic levels, irritability, and antibiotic needs for 6 months [74]. Bb12 is highly colonized because to its excellent adherence to human mucus [75]. The gut microbiome of preterm children has previously demonstrated a beneficial influence (**Table 4**) [70].

2.8.2.1 Bifidobacteria and colorectal cancer

Several research have looked at Bifidobacterium's ability to prevent and/or treat colorectal cancer. The bulk of research use mouse models to reach their conclusions,

Name deposit code	Proprietary company	Product	Example
Lactobacillus acidophilus NCFM ATCC SD5221.	DuPont Danisco	Heinz	nature Toddler
Bifidobacterium animalis subsp. Lactis BB12 DSM 15954.	Chr Hansen.	Heinz	Nestle Good Start
Lactobacillus rhamnosus GG ATCC 53103. Nutramigen	Valio	Mead	Johnson Nutrition
Lactobacillus reuteri DSM 17938.	BioGaia	Nestle	NAN L.I. GOLD

Table 4.
Probiotic strains of infant formula [76].

and the results imply that a combination of prebiotics and bifidobacterial may minimize the incidence of carcinogen-induced malignant cells in mice [77]. For example, it has been demonstrated that Bifidobacterium animalis has anti-mutagenic activity while growing in MRS broth, effectively counteracting the action of the carcinogen 2-amino-3-methylimidazo [4, 5-f] quinolone [78]. It has also been established in vivo and in vitro that a B. longum and a B. breve strain protect DNA from carcinogen-induced damage and suppress the genotoxic impact of two separate carcinogens when evaluated in a rat model [79].

2.8.2.2 Bifidobacterium and necrotizing enterocolitis

Following regular treatment of B. breve M-16 V, recent research found a decreased incidence of necrotizing enterocolitis in premature infants [80]. Administration of B. breve M-16 V in conjunction with breast-feeding was demonstrated to be related with a decreased incidence of necrotizing enterocolitis in neonates born before 34 weeks gestation, and, while not statistically significant, a lower incidence of this disorder was found for neonates born at a gestation age of less than 28 weeks [80].

2.8.2.3 Bifidobacterium and inflammatory bowel disease

Although the precise mechanism of action is unknown, probiotic strains were shown to reduce the symptoms of inflammatory bowel disease [81]. A probiotic mixture including three Bifidobacterium strains, four Lactobacillus strains, and one S. thermophilus strain was given to patients suffering from ulcerative colitis. Fifteen of the 20 patients stayed in remission throughout the experiment, indicating that treatment of this bacterial cocktail is useful in sustaining ulcerative colitis remission (**Table 5**) [81, 83].

2.8.2.4 Lactobacillus reuteri *strain*

L. reuteri is the probiotic strain of the probiotics *L. reuteri*, is a well-known probiotic. *L. reuteri* ATCC 55730 colonizes the stomach effectively and can reduce the occurrence of watery diarrhea associated with rotavirus. In addition, it recently showed effectiveness in the treatment of acute diarrhea with oral rehydration in

***Lactobacillus* species**	***Bifidobacterium* species**
Lactobacillus acidophilus	*Bifidobacterium adolescentis*
L. Casei	*Bifidobacterium animalis*
L. crispatus	*Bifidobacterium bifidum*
L. gallinarum	*B. breve*
Lactobacillus gasseri	*B. infantis*
Lactobacillus johnsonii	*Bifidobacterium lactis*
Lactobacillus paracasei	*Bifidobacterium longum*
L. plantarum	
Lactobacillus reuteri	
Lactobacillus rhamnosus	

Table 5.
Probiotic microorganisms [82].

children from 6 to 36 months of age [84]. A Indrio [85], research shows that *L. reuteri* DSM 17938 treatment has lowered stomach distension, expedited gastric emptying, and reduced regurgitation in children with normal gastroesophageal reflux. A recent study shows that this strain also has a favorable effect on the medication of baby colic if employed in several clinical studies or as a therapy for prophylaxis.

2.8.3 Saccharomyces *spp. with probiotic properties*

Saccharomyces genus contains several yeasts like: *Saccharomyces cerevisiae* that is used for the preparation of wine, bread, beer, *Saccharomyces bayanus* is used to produce wines, and *Saccharomyces boulardii* utilized in medicine as a probiotic [86].

S. boulardii is frequently advertised as a lyophilized probiotics to treat diarrhea and retains an excellent safety reputation [87]. Most reports show that *S. Boulardii's* clinical advantages are reducing diarrhea duration regardless of causation and thus reducing the social and economic benefits associated with hospitalization. *S. boulardii* dispensation has had a positive effect on the prevention and treatment of retroactive inflammatory bowel disease and moderate symptoms of ulcerative colitis [88] patients with irritable bowel syndrome. Current *Clostridium difficile* pseudomembranous colitis infection can also be drastically reduced through the administration of *S. boulardii* regular dosages together with standard antibiotics. No malformations were reported in the previously referred examination of probiotic safety during pregnancy [89]. It should be borne in mind, however, that *S. boulardii* can lead to fungal diseases or localized infections in immunocompromised people or in other patients.

2.8.3.1 Escherichia coli strains with probiotic properties

Although it is known mainly for its highly virulent serotype (e.g., *E. coli* O157:H7), *E. coli* is a very common lower gut inhabitant and even a probiotic strain is known to be *E. coli* Nissle 1917; *E. coli* genus belongs in a Gram-negative family called Enterobacteriaceae (EcN). As previously mentioned, it has been proven that constipation [90] and inflammatory bowel disease were treated with other probiotics in 1917 [91]. This strain could also alleviate gastrointestinal disorder, ulcerative colitis, Crohn's disease and even colon cancer.

2.9 Probiotic research as an advantageous facilitator in aquaculture

Researchers have previously shown that probiotic activities play a wide range in the host body (e.g., decreasing illnesses and stress, enhancing immunity, modulation of gut microbiota, nutritional assistance, improving quality of water, etc.). So, the positive effects of probiotics help to boost animal feed value and growth and improve aquaculture breeding and hatching rates. Probiotics have recently become a highly common technique in the aquaculture industry, and they are mostly isolated from fish guts. A recent study shows that Lactic acid bacteria (LAB), named Bifidobacterium, and Streptococcus are among the most common bacterial suggestions. Even though the use of probiotics in aquatic species is a relatively new concept, it has gotten a lot of interest because of its ability to influence many physiological processes.

In this study the many positive features of probiotics in aquaculture industries were proven. Probiotics are regarded as new functional agents with a potential impact on any aquatic organism's gut microbiome. Researchers have already shown that probiotic activities play a broad spectrum in the host body, such as reducing illnesses and stress, increasing immunity, modulating gut microbiota, nutrition aid, improving water quality, etc. In addition, the positive benefits of probiotics boost feed value and growth for the animal and improve the rate of aquaculture spawning and hatching.

2.10 Probiotics and their possible uses clinical importance

Probiotics are an interesting study field that the current age needs to examine for clinical wellness. Elite properties such as anti-pathogenic activity, anti-diabetics, anti-obesity, anti-inflammatory activities, anti-cancerous activities, anti-allergies and angiogenic effects and their influence on the intelligent and central nervous system (CNS) (**Figure 4**).

2.10.1 Probiotic anti-pathogenic action

Action of probiotics as anti-pathogenic is deemed as one of the most valuable effects in the probiotics, since the composition of the complex gut microbiota population is hindered, unlike conventional antibiotics, by disorder or change. Tejer in [93] examined the effect that probiotic substances can inhibit short-length fatty acid (SCFAs) pathogens (as acetic acids, propionic acids and lactic) on the survival activity of *Salmonella enterica*, Serovar typhimurium and *C. difficile* in a vitro scope model and postulated it. Kareem [94] investigated that SCFAs helps to maintain a colonic lumen pH that is imperative for the expression and for the metabolic rate of foreign compounds and carcinogenic substances in the gut [94]. Although it is known mainly for its highly virulent serotype (e.g., *E. coli* O157:H7), *E. coli* is a very common lower gut tenant and even a probiotic strain is known to be *E. coli* Nissle 1917; *E. coli* genus belongs in a Gram-negative family called Enterobacteriaceae (EcN). As previously mentioned, it has been proven that constipation [90] and inflammatory bowel disease were treated with other probiotics in 1917 [91]. This strain could also lessen gastrointestinal disorders like ulcerative colitis, Crohn's disease and even colon cancer.

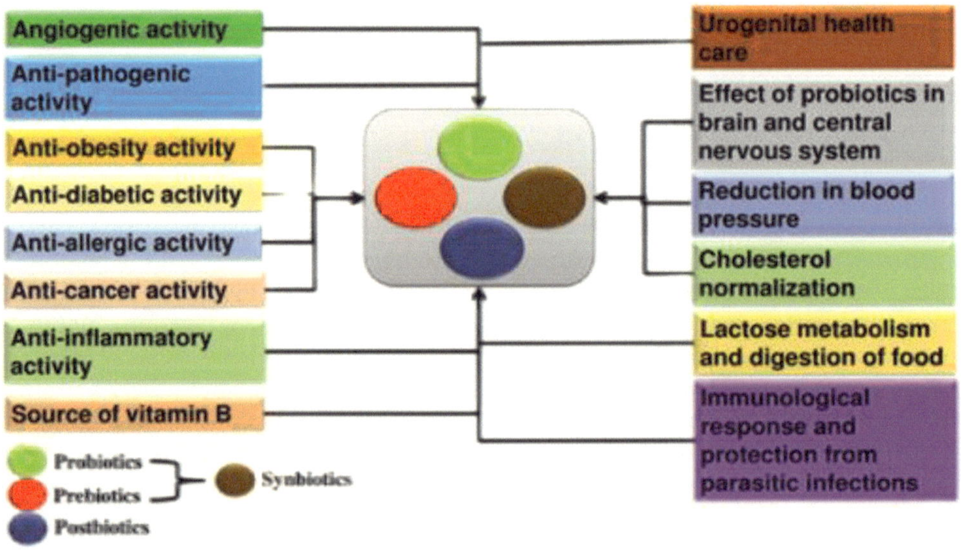

Figure 4.
Applications of probiotic [92].

2.10.2 *Anti-obesity activity of probiotics*

The physiological actions of probiotics are important for the health of the microorganisms controlling the environment of host. Thermogenic and lipolytic reactions in most cases assist loss of weight by activation of the sympathetic nervous system. *Lactobacillus gasseri* BNR17, probiotic strains, have demonstrated characteristics to block the rise and therefore restrict leptin secretion in adipocyte tissues, as their major source of leptin and adiponectin. Hypocholesterolemia effects have also been shown to be present in other probiotic bacteria, such as *L. casei, Lactobacillus acidophilus* and *B. longum.*

2.10.3 *The probiotics as angiogenic activity*

The term angiogenesis has proven crucial for the treatment of wounds and is needed to repair damaged tissue by delineating cellular responses [95]. The angiogenic programmed includes a set of cellular processes carefully regulated by which new vessels are created by the pre-existing cell reclamation and the production of cytokines, matrix-degrading enzymes, and chemokines. Angiogenesis that is deregulated is a key influence of cancer, diabetic retinopathy and IBD including CD and UC in main human illnesses [96]. Non-pathogenic *S. boulardii* probiotic yeast, protective from intestinal damage and inflammation, has been observed. However, these positive benefits remain unknown about the molecular mechanisms by which probiotics mediate. Probiotics may be potentially used to alter inflammatory cytokine profiles, decrease pro-inflammatory cascade regulation, induce regulatory mechanisms in a strain-specific way, strengthen the function of epithelial barriers, reduce the visceral hypersensitivity, increase traffic in spinal afferents, and reduce stress.

2.10.4 *Anti-allergic activity of probiotics*

In the past, probiotics have increased awareness of its causes and preventive measures in protecting and managing allergy disorders. In-vitro studies of some probiotic products, such as *Lactobacillus plantarum* L67, have demonstrated that the manufacture of interleukin 12 and interferon g at your host can prevent allergic diseases (park 2016). *L. plantarum* 06CC2 substantially relieved allergy symptoms in another research and decreased total immunoglobulin E concentrations, ovalbumin E specific immunoglobulin and histamine in the ovalbumin-sensitive mouse sera. *L. plantarum* 06CC2 is reported to enhance the interferon-g and interleukin-4 secretions substantially in the cells of spleen in mice, which alleviate allergy symptoms [97]. Further investigation may be useful in assessing the anti-allergic activity and method of action of probiotics.

2.10.5 *The probiotics as anti-cancerous*

On the World Health Organization's cancer data page, there were around 14 million new cancer diagnoses and approximately 8.2 million cancer-related deaths in 2012 alone. Asia, Africa, and the Americas account for more than 70% of cancer fatalities worldwide [98]. The attention has shifted to natural sources that impart anti-cancer benefits, such as probiotics, in recent years [99]. They are interested in working together to bring the illness down as well as produce a treatment with minimal or no adverse effects [100].

2.11 *In vitro* studies

The probiotic strains, the *Lactobacillus fermentum* NCIMB-5221, and -8829, have been shown to be extremely strong for the suppression and development of normal colonic cell epithelial growth by producing SCFAs in vitro studies (ferulic acid). Kahouli [101], in 2015 compared *L. acidophilus* ATCC 314 and *L. rhamnosus* ATCC 51303, that were characterized by tumorigenic activity. Probiotic strains of *L. acidophilus* LA102 and *L. casei* LC232 have shown the cytotoxic activities with two colorectal cell lines (Caco-2 and HRT-18) being in vitro anti-proliferative [102]. Although probiotics may play an important role in cancer neutralization, only *in-vitro* tests are confined to research. Therefore in vivo models and animals' clinical trials must prove the potential of anti-cancer probiotics.

2.12 Efficacy of probiotics in COVID 19

The recent Xu et al. [103] trials indicated the ability of probiotics in the avoidance of secondary infections in those afflicted by COVID 19. Some COVID-19 individuals suffered from microbial intestinal dysbiosis. All patients need to examine their dietary and gastrointestinal functioning. The regulation of the stability of gut microbiota and the decreasing probability of secondarily infected bacterial translocation should be supplemented with nutrition and application of probiotics.

2.12.1 Probiotics for COVID prevention

Probiotic medicines against viruses which lead to respiratory tract infections are proposed in the last two decades as antimicrobial agents. There are numerous conceivable action mechanisms to increase breath-probiotic activity; the modulation of the innate immune system and better immune response are nevertheless most probable. According to earlier research of certain viral illnesses, preventing infectious diseases can be achieved by increasing and activating human immunological activity by healthy, equilibrated meals and administrative complements such as vitamins, minerals, fiber, and probiotics [104].

Live microorganisms which provide a sufficient intake of health advantages, including an increase in immune activity and removal of respiratory tract diseases, are probiotics. Probiotics can obviously lower the prevalence and severity of illnesses, showing their promise to cure or prevent COVID-19. To manage viral infection, it is important to understand the immune cell activation, cytokine profile and immunological regulation. The preservation of the human GI or lung microbiota might help prevent COVID-19, as dysbiosis plays an essential role in people's vulnerability to infectious illnesses The potential preventative and therapeutic impact of probiotics against SARS-CoV-2 infection should be examined in in vitro and clinical investigations.

3. Isolation of probiotic microorganism

The initial process that was carried out for the isolation of probiotic bacteria is to keep the sample selectively before the step of incubation in suitable conditions. Many probiotics are anaerobic; hence, the samples then placed into anaerobic conditions immediately (within 3 h). The samples should be immediately homogenized, diluted, and cultivated in selective media Several mediums were developed to isolate bifidobacteria and lactobacilli either electively or selectively. As a source of isolation for probiotic bacteria were utilized milk fermented products (curd, buttermilk, cheese) and vegetable pickles. Direct plating and enrichment methods were used for the isolation of MRS agar and MRS broth, respectively. These samples were diluted in 9 ml of saline water serially up to 10-4 (0.89% NaCl) with a spread of semi-solid MRS on Petri flat plates. The inoculates were then incubated at room temperature for 2 days to examine and control microbial colonies that grown on medium (without inoculation).

3.1 Spreading

Appropriate sample dilutions were made, and a concentration of just 50 PI-J was dispersed on MRS-agar plates. For 48 h, each plate was kept at 37°C in a static incubator.

3.2 Streaking

A single bacterial colony was plucked with a sanitized loop and streaked over plates to get isolated colonies. The plates were then put in a static incubator at 37°C for 48 h.

3.3 Characterization of bacterial isolates

3.3.1 Morphological tests

These morphological tests were performed to identify bacterium isolates. These tests are as follows:

3.3.2 Gram's staining

This staining was used to discriminate between gram-positive and gram-negative microorganisms. Gram staining may be used to distinguish the morphology of bacteria, such as bacillus or coccus.

A neat and clean glass slide was prepared initially for gram's staining, and a thin smear of a single colony was created. The slide was then air dried. After that, the slip was fixed by running it through the flame 5–6 times. The prepared smear was then coated with crystal violet for 30–60 s. To remove any remaining discoloration from the slide, distilled water was utilized. For 30–60 s, Gram's iodine solution was applied to the smear. Following that, the initial stain was removed with alcohol for 30 s. The slide was drained with tap or distilled water before applying the secondary stain safranin for 1 min. Rinse the slide once again with distilled water. To dry the slide, blotting paper was employed. The slide was then examined under low and high magnification towards the end. Microscope power, i.e., at IOOX in oil immersion [105].

3.3.3 Endospore staining

- Endospore staining is used to distinguish spore-forming bacteria from non-spore-forming bacteria.

- To stain endospores, a crisp and clean slide was obtained, and a thin smear was created using an isolated culture of bacteria. The smear was then air dried. The smudge was fixed by passing it through the flame 5–6 times. Blotting paper was put across each slide after it had been fixated.

The malachite green stain was then put on blotting paper over steam for 15–20 min. After the slide had cooled to room temperature, the blotting paper was removed, and the slide was washed with distilled water for 30 s. The slide was then treated with safranin for about 2 min before being rinsed with deionized water. To dry the slide, blotting paper was used. Finally, the slide was examined under a microscope with low and high magnification powers, i.e., at 100× with oil immersion [105].

3.3.4 Motility test

A motility test is a test that is used to determine if bacteria are motile or not. Semi-solid medium was necessary for this purpose. Tryptone log, 5 g yeast extract, 13 g agar, and NaCl 5(g) were used to make the medium, which was then diluted in 1000 ml distilled water.

After that, 10 ml of it was poured into a variety of test tubes, cotton plugs were used to seal the test tubes' mouths, and it was sterilized in an autoclave at 121°C for 15 min. Finally, the medium was allowed to harden vertically. The medium was infected using a red-hot inoculating needle and then incubated at 37°C for 24 h [105].

3.3.5 Biochemical characterization

3.3.6 Catalase test

This test was carried out to determine the capacity of microorganisms to digest hydrogen peroxide (1-1202). A nice and clean glass slide was used to perform the catalase test. In the center of the slide, one drop of water was placed. Using an inoculating loop, I took some isolate culture and mixed it with water. I applied 2 drops of hydrogen peroxide to it and witnessed the results [105].

3.3.7 Casein hydrolysis test

Skim milk agar medium was made for the casein test by combining 2 g tryptone, 1 g yeast extract, 6 g agar, 4 g glucose, and 4 g skim milk in 400 ml of distilled water. The media was steam sterilized in an autoclave for about 15 min at 121°C, and the petri plates were poured under sterilized conditions. The isolated colony of each bacterium was then streaked in the middle of the petri dish under sterilized conditions. Each plate was placed in the incubator at 37°C for 24 h before the change was detected in each petri dish [105].

3.3.8 Carbohydrate fermentation test

The goal of this test is to confirm the microorganisms' ability to ferment carbohydrates via gas and acid [105].

3.3.9 Glucose fermentation

The medium for this experiment was phenol red broth. 5 g NaCl, 0.018 g phenol red, 10 g peptone, and 5 g glucose were dissolved in 1 liter of distilled water to make phenol red broth. The medium was then autoclaved for 15 min. The bacterial culture was then injected in the medium. For 24 h, test tubes were put in an incubator set to 37°C. If the color goes from red to yellow, the result is good. To observe gas generation, a Durham tube was inserted in each test tube [105].

3.3.10 Lactose fermentation

The medium for the lactose fermentation test was made by combining 0.018 g phenol red, 5 g sodium chloride, and 10 g peptone 5-gram lactose in 1000 milliliter deionized water. The medium was then autoclaved sterilized. The medium was then injected with a bacterial culture. After that, each test tube was placed in the static incubator overnight at 37°C. The presence of yellow suggests a favorable outcome. Durham's tube was used to monitor gas output.

3.3.11 Sucrose fermentation

The sucrose fermentation test was performed to determine whether or not a microorganism had the capacity to ferment sugar. To conduct this test, the following substances were dissolved in 1000 ml of distilled water: 10 g peptone, 0.018 g phenol red, 5 g sodium chloride, and 5 g sucrose. The medium was then sterilized in an autoclave at 121°C for 15 min. After that, the medium was injected with bacterial culture. For 24 h, the test tubes were put in an incubator set to 37°C. If a yellow tint develops, it indicates that fermentation is taking place. Durham tubes were inserted in each test tube to observe gas generation [105].

3.3.12 Probiotic properties of isolates

The following are the primary selection criteria for determining the probiotic characteristics of bacterial isolates.

3.3.12.1 NaCl tolerance test

Bacterial isolates were exposed to various NaCl concentrations to determine their tolerance range. For this, MRS broth containing 1–4% NaCl was injected with 0.1 ml of each bacterial suspension's stimulated bacterial culture and incubated at 37°C for 48 h. The growth rate of the organism was estimated by obtaining the O.D. in a spectrophotometer at 600 nm.

3.3.12.2 Antibiotic sensitivity test

This experiment was carried out to determine antibiotic sensitivity. Antibiotics such as amoxicillin, penicillin, erythromycin, ciprofloxacin, gentamicin, and cefixime were used for this. First, the freshly cultured bacterial isolates were dispersed over the sterile filled Nutrient Agar plates. The antibiotic discs were then put on petri plates at identical distances and incubated at 37°C for 18 h. The creation of a clean zone surrounding the discs was seen after 18 h.

3.3.13 Molecular characterization of bacterial isolates

The following steps were used to characterize isolated bacterial isolates molecularly.

3.3.13.1 Isolation of genomic DNA

In 20 ml of autoclaved LB broth, isolated colonies of chosen bacteria were inoculated. For 18 h, these falcons were kept at 37°C in an incubator that was constantly shaken. After incubating for 18 h, the broth culture was put into an Eppendorf tube. That Eppendorf was centrifuged for 5 min at 40°C and 6500 rpm. The particle was kept, but the supernatant was discarded. Pellet was thoroughly washed with 200 PI of TEN buffer and mixed with a vortex.

This combination has been centrifuged for 5 min at 40°C and 6500 rpm. The supernatant was discarded once more. The pellets were preserved for future use. Following that, 100,111 of SET buffer was put into an Eppendorf and the pellet was well mixed using a vortex. The Eppendorf was then filled with 100 gl lysozyme and placed in the incubator at 37°C for 30 min.

Following that, 5 PI of 25% SDS solution and 100 PI of TEN buffers were added. The Eppendorf tube was gently inverted many times until lysis occurred, and then incubated at 600°C for 15 min. After withdrawing the Eppendorf from the incubator, it was allowed to cool at ambient temperature before being filled with 5 gl of 5 M NaCl. The mixture was treated with an equal proportion of chloroform and buffered phenol (l:l). Eppendorf was centrifuged at 40°C for 1 min at a speed of 6500 rpm. Supernatant was removed once more and transferred to a fresh Eppendorf tube. The DNA was then precipitated by adding twice the volume of absolute ethanol (100%) that had to be ice cold. Overnight, the Eppendorf was chilled.

The next day, Eppendorf was centrifuged for 5 min at 6500 rpm (40°C). The supernatant had been decanted. The pellet was rinsed with 70% ethanol. The Eppendorf was properly air dried. TE buffer (50 microliter) was added, and the DNA was kept at 20°C for future use.

3.3.13.2 Gel electrophoresis

Gel electrophoresis was used to determine if the treated materials contained isolated genomic DNA or not. 1% agarose gel was made for this purpose by dissolving 1 g of powdered agarose in 2% 50× TAE buffer. 2 ml of 50× TAE buffer was dissolved in 98 ml of autoclaved distilled water to get the 2% solution. Unless the agarose and buffer solution was correctly mixed, it was heated in the microwave for over 2 min. After cooling to room temperature, ethidium bromide 2 PI was added. When the temperature was reduced to 600°C, the gel was gently poured into the gel casting tray. Before pouring the gel, the comb was placed in the gel casting tray. The gel plate was put on a level surface to make the gel smooth and consistent. A 20 PI sample was obtained in an Eppendorf tube and 5 gl of 1× loading dye was applied to it. The sample was stored on ice during processing. After solidification, the comb was carefully removed from the tray, and the test sample was placed in each well. For over 40 min, electrophoresis was performed at 80 volts. A DNA ladder (10 Kb) was put into one well.

3.3.13.3 Polymerase chain reaction

Ferment was used as a PCR reagent in the PCR of 16S rDNA. The samples were put into a Thermo cycler (Progene, Techne) that was set for initial denaturation at 940°C for 20 min, melting at 940°C and annealing at 5200, primer extension at 720°C for 60 s each, for a total of 35 PCR cycles (**Figure 5**).

The last extension at 720°C lasted 10 min, and the ultimate storage temperature was set to 40°C for the maximum term.

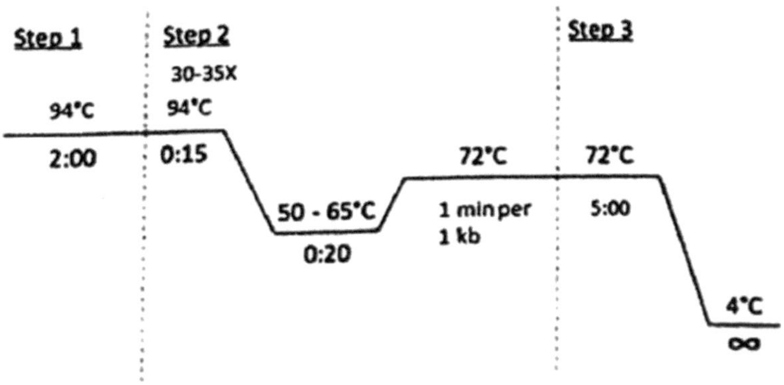

Figure 5.
PCR steps.

3.3.13.4 Gel electrophoresis of amplified product

1% agarose gel was created for this purpose by combining 1 g of powdered agarose with 2% of 50× TAE buffer. Amplified items were placed in the appropriate wells. It was ran at 80 volts for 40 min. The gel was then examined for the presence of amplified products using a UV trans-illuminator.

3.3.13.5 Purification of PCR product from agarose gel

After amplification, the required DNA band was sliced, and the net weight of the agarose gel containing the DNA band was calculated. After that, an equivalent volume of binding buffer was added and the gel was incubated at 50°C until fully melted. The sample was then transferred to a column. The column was centrifuged at 10,000 rpm for 1 min, then washed with 750 gl of wash buffer and centrifuged again for 1 min. The flow through was removed, and the column was washed one more with 750 gl of was buffer. For 1 min, the column was centrifuged at 10,000 rpm. Flow through was discarded once again, the column was moved to a fresh micro centrifuge tube, and 30–50 PI elution buffer was added. It is then permissible to centrifuge for 2 min at 10,000 rpm for 30 s and store at −200°C for future use.

3.3.13.6 Sequencing

Following purification of the PCR products with the-Invitrogen gene clean, the samples were forwarded to the laboratory for 16S rRNA sequencing.

3.4 Identification

The initial stage in the identification of potential probiotics is to identify microorganisms inside the GIT or in dietary sources. Only a tiny proportion of microorganisms can be cultivated in various habitats in culture [106]. The taxonomic categorization characterized as the process of cataloging that was based on a polyphasic approach [107]. Phenotypical techniques used to identify microorganisms have historically been employed. For many decades, the taxonomy depends on significantly on the kind of sugar fermentation and fermentation products. The probiotics were therefore categorized largely as LAB. The method of choice today is 16S rRNA gene analysis. Microbiologists have employed this conserved region for phylogenetic categorization for the past two decades, and the relatedness of species is inferred by comparing their sequences in publicly available databases. To detect bacterial communities from gutorecological sources, 16S rRNA gene analysis was coupled with other techniques.

The amplified 16S rDNA may relate to PAGE by means of the hybridization using fluorescent oligonucleotide probes (fluorescence in situ hybridization) or chemical denaturation using restricting enzymes (T-RFLP) with a specific 16S. However, in comparison with the bacterial Genome having base pair of 30,000–40,000, the 16S rDNA segment is exceedingly tiny (1500 bp).

3.5 Characterization

The two most significant probiotic-taxa are *Lactobacillus* and *Bifidobacterium* species in processed dairy products. When eaten, sufficient metabolically active bacteria are required to penetrate the GIT barrier and have transitory impact in GIT. This is essential since some writers have demonstrated the positive benefits of dead probiotics [108]. GIT has the challenge to survive on GIT with the potential to withstand with extremely low pH of about 1.5, availability of gastric enzymes, bile salts and other bowel enzymes [109]. Different in vitro tests to imitate these stress conditions have been devised.

3.6 Identification and characterization of probiotics

Isolate were identified by gram staining, endospore stain, catalase testing, and carbohydrate fermentation test. The growth and survivability of the stomach and small intestines is part of this. The stability of these properties following ingestion must thus be tested to verify that they are maintained in the host. Therefore, tests of acid and bile tolerance should include early screening and selection of probiotic strains.

Classical physiological and biochemical assays are not efficient for analyzing and quickly identifying microbial communities, as the bacterial population typically has comparable nutrient requirements and develops under similar environmental circumstances. Thus, it may often be difficult to clearly identify the species using simple phenotypic criteria. New possibilities for defining strains of fermented milk items have been developed using molecular methods. The 16S rDNA assays are fast, and cheap to detect the microbial species of yeast, acetic acid, and of some Gram-positive bacteria. Among PCR-based techniques is easy and cost effective. The chosen strain for diverse probiotic characteristics was characterized. These comprise the susceptibility analysis of antibiotics, the capacity to create bioactive metabolites and acid sensitivity tests for their antibiotically resistance potential.

3.6.1 Antibiotic susceptibility test

Several antibiotics were employed at different concentrations, ranging from 25 μg, 50 μg, 100 μg, 200 μg, 250 μg and 500 μg/ml agar medium, including the penicillin G, tetracyclines, gentamycin, vancomycin, and streptomycin. Lactobacillus has been streaked across an agar plate over the overnight culture. It was aerobically incubated at period of 48 h for 37°C, controlled and checked for lack of growth.

3.6.2 Antimicrobial production

Bilkova et al. have evaluated the antimicrobial production of the chosen isolate. As an indicator strain E. coli was utilized. The supernatant culture was caught at the time period of different intervals consist of 21, 24, 27, 30 h of inoculation following intervals. Above supernatant the proteinase, followed by heat treatment, was centrifuged, and digested. It was put into 200 μl wells in E. coli lawn containing plates and incubated with control (sterilized water) for 37°C for 48 h to check for the presence of the indicator strain growth inhibition zone.

3.6.3 Acid sensitivity test

In MRS broth it was controlled by examining the capacity of isolated bacteria to grow at the acidic environment in different pH. After 24 h incubation at 37°C at 150 rpm, OD values at 600 nm were measured. The bacterial growth was compared with growth in the pH 7 MRS broth by determining its acid sensitivity.

3.7 Studies demonstrating characterization and identification of probiotic isolates

3.7.1 Isolation from cheese sample

Ward and Timmins [110] published another investigation which found that three strains isolated from cheese samples having *Lactobacillus casei*, *Lactobacillus paracasei* and *Lactobacillus rhamnosus* were distinguished by PCR. In MRS agar samples, 63 isolates were cultivated and evaluated with sugar fermentation phenotypes. These isolates then belonged to the casei group. PCR primers were developed for 16S rRNA gene preserved areas and particular fragments were amplified with PCR. The agarose gel separated the amplified products from the data base and retrieved their gene sequences. These three distinct strains in the region VI of 16S rRNA have given unique amplified products. At the end of the process, a product containing the 16S rRNA *Lt. paracasei* gene was discovered in 51 of 63 isolates and a product containing the 16S rRNA gene *L rhamnosus* was given in 12 of 63 by primers, whereas a product containing the *L. casei* primers was not detected.

3.7.2 Bacterial viability protection during in vitro *gastric transit simulation in fresh cheese from Argentina*

Vinderola et al. [111] examined fresh cheese from Argentina a type of soft rindless cheese that have a 12-day maturation at 5°C before it is sold on the market. The following parameters are shown in this cheese: 5.29 pH, 58% humidity (w/w), 12% fat (w/w), 23% proteins (w/w), 0.9% salt (w/k), ashes (w/w), 40.8% dry matter (w/w) and 0.6% calcium (w/w). This product has shown suitable for the preservation and consumption of probiotic microorganisms. It gives a certain level of bacterial viability protection during in vitro gastric transit simulation.

3.7.3 Counts of L. acidophilus *to improve the flavor, and texture*

Kasimoglu et al. [112] demonstrated that the strain of *L. acidophilus* may be utilized in the production of the probiotic Turkish white cheese. To make health claims, the finalize counts of *L. acidophilus* were more than the minimum (107 Cfu g^{-1}) needed. *L. acidophilus* may also be employed in the production of high levels of proteolysis to improve the flavor, and texture. In addition, vacuum-packed probiotic cheese was demonstrated to be more acceptable after salting than the same cheese kept in salt. The recommended way of preserving probiotic Turkish white cheeses is thus the vacuum packing.

3.7.4 Probiotics in cheddar cheese

Phillips [113] researched about probiotic cheddar cheese. Six samples of this Cheese were produced with various mixtures of probiotic flora that was commercially available. Every supply, the specie of *Bifidobacterium, Lactobacillus acidophilus* and either *Lactobacillus casei, paracasei, or rhamnosus* have been present in cheeses. They described adequate viable counts and a beneficial impact on the consistency and sensory characteristics of cheeses. Cottage cheese has a suitable profile for the inclusion of probiotic cells. Moreover, cottage cheeses are, due to their low-fat content, a healthier alternative to many other cheeses.

3.8 Isolation of bacterial strains done by marketed foods and drugs

Liu et al. [114] assesses 41 lactic acid strains, 36 of which have been identified and obtained from the commercial produced milk and pharmaceutical products, and 5 forms of probiotics were assesses that were obtained from China. These samples then incubate in the incubator for the period of 2 days in the conditions of anaerobically provided medium at the temperature of 37°, so different colonies have been morphologically chosen and that were classified as rods or cocci under a light microscope. After being plated on suitable agar plates, pure colonies have been isolated. The conventional microbiological techniques of colony appearance, gram staining, oxidase, and catalase reactions were initially used for all isolated isolates. Amplification of the PCR and further sequencing of 16S rDNA at genus level have been carried out. For identification purposes, universal primers 27F and 1492RP were employed. AGAROSE Gel was then separated and purified by electrophoresis by 1.5% (w/v). ABI DNA Sequencer 3730 was used to the purified products. The alignment of 16S rDNA was done by BLAST.

3.9 Isolation of probiotics from milk and fermented derivatives

Mishra and Sharma [115] isolated probiotics from milk and its fermented derivatives such as buttermilk, curd, and cheese. Following direct plating on MRS agar, the colonies were serially diluted in saline water (0.89% NaCl) up to 10-4 before being distributed over MRS agar plates and cultured for 48 h. Purified isolate colonies were streaked on agar plates and morphological features of colonies were used to identify them. Gram staining, catalase test, and carbohydrate fermentation test were used for physiological characterization, and the results revealed five types of isolates, with bacteria isolated from curd and buttermilk samples being gram positive, catalase negative, and capable of fermenting glucose and mannitol without manufacturing gas. The isolates from the cheese and milk samples were determined to be gram negative, thus they were not included in the study. The isolates from the curd sample were further examined for antibacterial activity and antibiotic susceptibility, with positive findings indicating that they belonged to the *Lactobacillus casei* genus.

Lactobacilli isolated from traditional milk products of 17 sample samples known in Azerbaijan as tvorog curd cheese. 17 samples have been taken of tvorog, followed by 1 g suspension of each sample in solution of saline. The MRS broth was then added 500 μl of suspension and then it was incubated for the period of 48 h at 37°C. In the concentration of 10.0 ml PBS buffer (pH = 3), 1.0 ml of each cultivation of the enriched culture was incubated for 3 h to detect lactobacilli resistance to severe stomach conditions. The 10 ml of broth was added and incubating at 37°C for the period of 4 h resuscitated pH-resistant bacteria after centrifugation. Another test was performed to determine the bile salt resistance for bacteria. The bacteria that were acid resistant were injected in MRS broth and incubated for 4 h at 37°C. The dilutions were placed on MRS agar plates and incubated for the period of 24–48 h at 37°C. In 10 ml MRS broth, different colonies were selected and cultivated. The isolates were first assessed using gram staining and the morphology of cell. After that, the isolates were stored at −70°C in MRS broth with 10% skim milk and 25% glycerol. All these species have antibacterial properties against the indicator bacterium. Isolation of antibacterial samples by tvorog curd cheese, to determine acid and bile resistant lactobacilli strains, that was identified by 16 s rDNA as *L. plantarum, L. casei, and L. rhomnos.*

Tavakoli [116] gathered five specimens of Koozeh from remote area of Mazandaran province. The sample (30 g) has been homogenized and then incubated for the period of 24 h at 37°C and inoculated in 300 ml of MRS Broth. In the following stage, the pellet was resuspended to be 10 ml phosphate-buffered (PBS) in a 2.5 pH-adjusted phosphate-buffered Saline (PPS), then placed on an MRS agar medium. The plate was incubated in anaerobic circumstances at 37°C for 24–48 h. Colonies of all morphologies were gathered and purified by culture on the same medium. Gram positive, catalase negative isolates were regarded as a presumptive LAB following gramme staining and catalase reactions, which was kept at −80°C in 15% glycerol. The different isolates were examined based on colony morphology (form, superficie and color), cell morphology (form and size), and the biochemical properties of phenotypes and lactobacillus. They were tested. Jayne Williams in 1977 described or explained that the genus Lactobacillus that depends on the morphological state, and biochemical features, has been deemed to include eight bacterial isolates. For the sequencing, the typical 16S rDNA amplicons were picked for each of the several profiles. The homology of the sequence sequences was more than 95% for four distinct Lactic acid bacteria species [117] in terms of molecular identity comparisons. There, *L. plantarum, L. casei, L. pentosus, and L. fermentum* were as follows.

3.10 Probiotic potential of LAB isolates

This study examined the propensity to promote the activity of β-galactosidase, CSH, generation of hydrogen peroxide, antibiotic sensitivity, and pathogenic microbes, in vitro, of eight strains of lactobacillus isolated from Iranian conventional cheese, "koozeh," The results demonstrate that all lactic strains are powerful probiotics in developing novel formulations for the design of health-promoting functional food items. Analysis showed that the best of test probiotics among the tested are *L. fermentum* named MT. ZH893 and MT. ZH993 and MT. ZH593 *L. plantarum. Lactobacillus acidophilus* is an initially isolated bacterial bacterium economically important and was named *Bacillus acidophilus* from human gastrointestinal tract in 1900. During the development of technologies for the identification of bacteria *L acidophilus* is the typical species of a very varied and heterogeneous Lactobacillus group that has undergone several taxonomical revisions. The characterization of

L. acidophilus has suffered with misrepresentation because of the difficulties of distinguishing phenotypically identical species by morphologic and different biochemical techniques. In comparison, *L. acidophilus* sensu stricto is currently one of the most characterized species of *Lactobacillus* and it is used as a supplement for probiotic that is present in functional foods. The source of *L. Acidophilus* strains is established, *L. acidophilus* historically and now misidentified, and the probiotic has genomic, and physiological features.

Another study conducted by Maged in [118] identified 93 lactic acid bacteria (LAB) from the local and fermented milk samples of 13 in number that were gathered, including fresh raw milk, crude frozen milk, and different types of cheese like (fresh, salty, cooked) yogurt stirred, (shrimp milk) yogurt, and butter. The LAB counts were greater under circumstances of microaerobic incubation than in conditions of aerobics. On the MRS solid medium surface, the colony morphologies of the isolates were seen; the colors of colonies observed were white to pale creamy with circular shape and the width that ranged from a diameter of 0.5–4 mm. About 92.47% strains were gram-positive. Catalase and cytochrome oxidase activities varied in isolated isolates, as seen that 71% were unfavorable for the activity of catalase and cytochrome oxidase production were negligible about 72%. Also, 96% of isolates were nonmotile and the motility of isolated strains varied. All the isolated strains were amplified with the 16S rDNA gene. Then by the usage of amplicon it was being sequenced and purified. The 46 distinct genera's nucleotide sequence, i.e., *Enterococcus, Lactococcus, Lactobacillus, Streptococcus* and *Weissella*, are in line with 16S rDNA sequences from 14 species. The nine strains of the genus *L. acidophilus, L. casei, L. paracasei, L. plantaarum* and *L. futsaii* were analyzed to indicate that there could be two strains named (i.e., Hadhramaut4 and Musallam2), four strains named (i.e. MSJ 1, BgShn3, MasaLam7, and Dwan5), one strain namely (i.e. NMBM1), and another strain, (i.e., EMBM2), respectively.

Ten condensed yogurts imported from China were studied by Qian [119]. In the MRS medium, the strains of *Lactobacillus* grew at a temperature of 37°C at 16–24 h. Brain Heart Infusion Gardnerella vaginalis (ATCC49145) with yeast extract supplement (1%), maltose (0.1%), glucose (0.1%), and horse serum (10%) (BHIS) at 37°C for 24 h under anaerobic circumstances. It was cultured on Luria- Bertani Medium (LB) for 12 h at a temperature of 37°C (ATCC 25922). Streaking is done by MRS agar and broth and enriching them for the strains of *Lactobacillus*. Gram-positive colonization's were chosen and inoculated in a broth of MRS with characteristic *Lactobacillus* shape, white in color and fruity fragrance. Then genetic investigation with the use of PCR and 16S rDNA sequencing has validated and identified isolates. A bacterial DNA isolation kit was used to extract the genomic DNA from the strains of *Lactobacillus*. The 16S rDNA genes were amplified using the universal PCR primers 27F (AGAGTTGATCGGCTCAG) and 1492R (TACGGC TACCTTGTAGACTT).

3.11 Study of phylogenetic analysis of probiotics

Hajigholizadeh et al. [120] isolating LAB from traditional cheeses and characterizing them. In 225 ml the quantity of peptone water is about 0.1% w/v was added and then mix the 25 g of each sample of cheese. Then dilute the sample of cheese in a suspension containing 2% of the sodium citrate and then grown on MRS agars and incubated for 1–2 days at a temperature of 37°C under anaerobic and aerobic conditions. From each plate of cultivation, the 3–4 distinct colonies were picked randomly. Gram staining microscopically inspected, and the catalase analysis was carried out.

Molecular and antibacterial characterization is analyzed and kept in a test tube that contain 15% (v/v) glycerol at −20°C. Extraction and amplification of bacterial genomic DNA. Different kinds of bacteria colonies emerged on MRS agar surface following screening and phenotypic characterization. Screenings were performed on different 60 MRS agar plates that have tiny, round, matt, and white colonies, it was performed with 70 different biochemical characteristics of LAB showcasing the gram-positive and the catalase negative, including cocci, or in shape rods.

3.11.1 PCR and RFLPs

Koohestani [121] study based upon the spot-on-the-lawn approach was performed to assess the performance regarding antibacterial activity of 8 LAB isolates with various patterns of RFLP. For all 70 bacterial isolates, a DNA fragment contains the size of 1.540 bp has been enlarged as shown in **Figure 6**. Three distinct digestive patterns were shown in the RFLP PCR product analysis (patterns I–III) shown in **Figure 6**. Out of 70 isolates, RFLP pattern I was shown in amplified fragments from 16 rRNA from

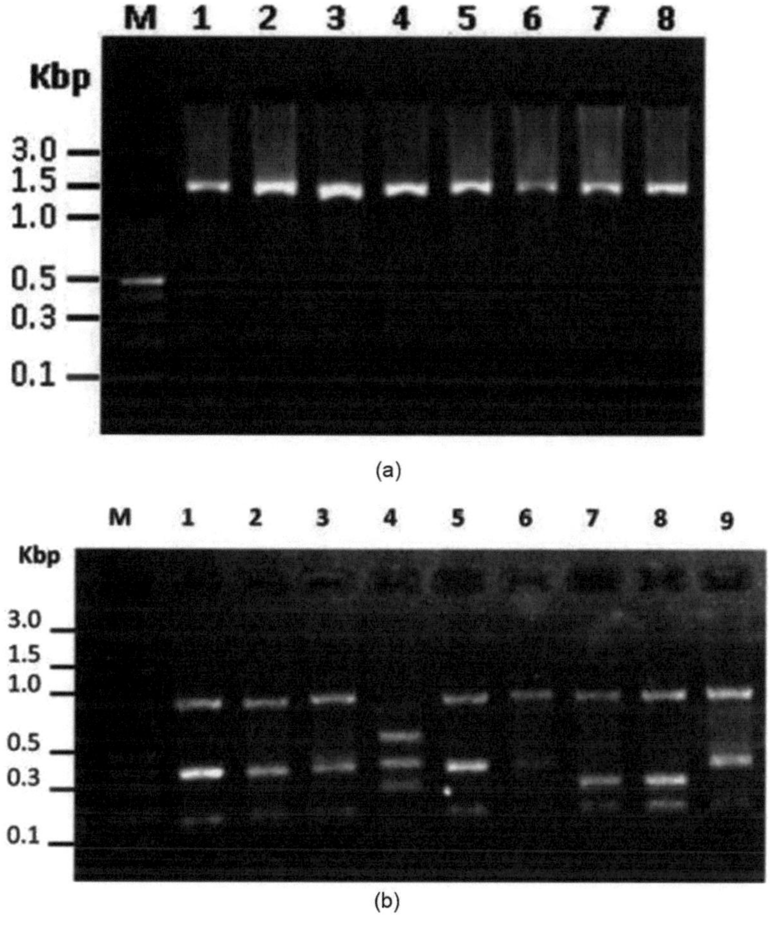

Figure 6.
Staphylococcus planktonic form and biofilm were inhibited by cell free supernatant of Lactobacillus casei [121].

63 (90%) isolates. The RFLP patterns I and III were all RFLP patterns discovered with the HinfI named endonuclease enzyme, and each with two and five isolates.

LAB isolates were identified based on the generation of the phylogenetic trees from the sample of cheese named *Enterococcus* subsp., *Lb. lactis*, *Lb. farciminis*, and *Lb. paracasei Enterococcus* subspecies was made up of majority of the LAB (90%) in that RFLPs isolates named [2, 97, 122, 123] isolate it. RFLP pattern II was classified jointly in Isolates namely 14 and 32. The RFLP pattern III was identical to two isolates 22 and 44; nevertheless, these two isolates were grouped into two different clusters based on a phylogenetic tree. The most common LAB in traditional cheeses in this research were *enterococcus* subsp.

3.12 *Enterococcus* safety evaluation and probiotic potential by genomic analysis

Enterococci are the ordinary people of the human and animal gastrointestinal system. Recently, the selection of helpful microorganisms by bacteriocins was a novel probiotic characteristic [124]. The 1st OB14 strain and 2nd OB15 strain of lactic acid were isolated and identified as *E. faecalis* from Testori and Rigouta typical Tunisian fermented milk products. These novel isolates have been examined for the character of the gastrointestinal tract and proven tolerant to severe circumstances. They were moderate biofilm makers that increase the trans epithelial resistance and they can attach to the cells of intestines to reinforce the barrier. Different antibiotics susceptibility is seen includes Ampicillin, vancomycin, gentamicin, and erythromycin and the evidence shows the susceptibility of *E. faecalis* OB14 and for OB15 to essential ampicillin and vancomycin clinical drugs. The tetracycline resistance existence and the presence of cytolysin genes in *E. faecalis* OB14 1st strain, nevertheless, was found in the Whole Genome Sequencing (WGS). Hierarchical cluster analysis reveals the tight connection between *E. faecalis* OB15 and *E. faecalis* Symbioflor 1 against OB14. *E. faecalis* OB15 looks therefore trustworthy as a probiotic food or feed business for future growth.

3.13 Study of strain of probiotic potential by isolation and identification

Samples of Ezine type of cheese have been mixed and attenuated into the solution of Ringer and plated with aerobic incubation at 37°C at a species of kanamycin aesculin azide agar for 48 h. After incubation, colonies exhibiting Enterococcus-typical shape were randomly chosen and spotted using sterile toothpicks on plates containing agar. A total of 114 colonies with characteristic Enterococcal shape were transferred to BHI Agar in KAA agar. It was observed that 84 colony areas greater than 10 mm versus indicator strains displayed inhibitory zones (results not shown). PMD74 was identified as the strongest antibacterial activity for additional tests. Among these colonies. The distinctive characteristics of the strain are compatible with general characteristics. The isolate has been identified as an Enterococcus lactis strain, according to the sequence of 16S rRNA, undertaken to ensure molecular identification.

The latest investigation showed that Ezine (PDO), which consists of nonstarter Turkish white long-ripened cheese, serves as an isolation source for new enterococcal strains. This is the first analysis on *E. lactis* isolation in Turkey, to the best of our understanding. *E. Lactis* is a probiotic candidate because of the results such as strong strain tolerance to GI tract virtual circumstances, other physiological features, and remarkable antibacterial activity to both near relatives and dietary-borne pathogens.

3.14 Commercial interest in probiotics and sensory evaluation of food matrix

Commercial interests also exist for the idea of probiotic food, as is shown in the range of probiotic products accessible in supermarkets and specialized stores, that make up a major portion of the functional food market. Singh (2011) observed that various author has demonstrated that frequent ingestion of live probiotic microbes might be useful to improve lactose tolerability, reduce cholesterol levels. It was observed that probiotics may directly or indirectly affect the intestines by modifying the physiology of Endogenous Microflora or Immune System, as colonization of some strains can decrease the severity of acute diarrhea in children. Probiotics have a positive impact on the intestinal microbiology, including antagonistic effects, competition for immunological effects and improved infections resistance.

The usage of bacteria at the expense of potentially dangerous bacterial proliferation therefore encourages the proliferation of beneficial bacteria which enhances the natural defensive systems of the host. In fermented milk products, the application of probiotic bacteria has been extensively explored due to problems in maintaining the vitality of these organisms during cooling storage. The survival of probiotic bacteria in fermented dairy product may be influenced by factors such as acidity and dissolved oxygen and species interactions, inoculation techniques and stock conditions.

The quantity of viable bacteria in the intestines and the level of pH that is low in stomach leads to the limit the survival of probiotics. Furthermore, there are still numerous difficulties with the poor viability of probiotic bacteria in milk meals. In fermented dairy products there are several variables that impact the survivability of probiotics: acidity, pH and hydrogen peroxide, dissolved concentrations, stock-temperature, interaction in products with other microorganisms, lactic and acetic acid concentration, and protein concentration [125].

4. Some highlights on the LAB and possibilities for the future

Without initially securing food safety, there can be no feeling of global security. This involves, among others, supplying the world's thriving population with safe and healthy food. The future is undoubtedly hopeful given the tremendous potential in the utilization of LABs as probiotics. One topic now being explored, for example, is the examination using whole genome sequence technology of probiotic propensities. Among other effects, the functions of probiotic LABs will be improved, and data may be utilized to further modify LAB genes [126]. More study is being undertaken on their usage as functional food components and will expand soon. The effects of probiotic LABs on the cells of breast and likes to have previously been researched to bridge the gap between the world's food, medicinal and health industries. However, it must be noted that the safety for the improvement of food is not a probiotic feature of lactic acid bacteria (LAB).

Over the recent decade, numerous researches have been carried out examining the molecular foundation for possible probiotic characteristics of prospective LAB strains and their products, drastically enhancing our biological understanding [127]. These findings have been and still can form the foundation for important in vitro and in vivo investigations for food, biomedical and pharmaceutical specialists. These studies are crucial. Preliminary findings from current study at the Northeastern Agricultural University's Key Laboratory of Dairy Science (KLDS).

Based on the research provided in this review, various strain combinations may thus be hypothetically utilized to evaluate the attenuation effects on gut-microbiota with respect to obesity and T2D of probiotic LABs and *Bifidobacterium* species. To create new anti-obesity foods and dairy products, similar research evaluating the effect of various strains on the gut microbiota may be carried out. Such research might open fresh and new medicinal and food goods pipelines with huge industrial uses.

5. Conclusion

Probiotics mean live organisms that have positive effects on the health of host. The aim of current study was to isolate probiotic bacteria from different non-dairy products. The total of 1 1 samples was collected from different areas and from these samples 10 bacterial strains were isolated. All of these were characterized based on morphological, biochemical tests and 16S rRNA ribotyping. Probiotics are helpful bacteria that are catalase negative and help to restore microbial equilibrium in both people and animals' guts. Lactobacillus was the first probiotic to show a positive influence on health. These "Good Microorganisms" can be gotten not only from non-dairy products, but also from processed dairy products such as. They improve epithelial barriers, increase adherence to gut mucosa and microbial adhesion, produce antimicrobial compounds, and regulate immune systems, and can be used as food supplements to treat various gastrointestinal tract diseases, as well as in research studies to develop commercial probiotic foods. Probiotics have health advantages such as improved immune responses, easing of lactose intolerance symptoms, diarrhea treatment, cholesterol decrease in serum, vitamin generation, and anticarcinogenic properties. Probiotics serve a variety of roles in the host body (e.g., decreasing illnesses and stress, enhancing immunity, modulation of gut microbiota, nutritional assistance, improving quality of water, etc.). As a result, the beneficial effects of probiotics contribute to increased animal feed value and growth, as well as improved aquaculture breeding and hatching rates. Probiotics have the potential to treat or prevent COVID-19 by lowering the occurrence and severity of diseases. *Lactobacillus casei* also interacts with epithelial cells via Toll-like receptors (TLRs) to boost the production of cytokines, which are vital in increasing cell productivity and preventing apoptosis during restoration, promoting survival and proliferation. The preservation of the human GI or lung microbiota may aid in the prevention of COVID-19, since dysbiosis plays an important role in people's susceptibility to infectious diseases. Most experimental experiments demonstrated that bacteria extracted from processed dairy products belonged to lactic acid bacteria, which are classified as probiotic bacteria. Appropriate sample dilutions were made, and a concentration of just 50 PI-J was dispersed on MRS-agar plates. For 48 h, each plate was kept at 37°C in a static incubator and all the steps were done for performing the tests. Then electrophoresis and pcr is done to confirm the antimicrobial activity.

Various research findings on the importance of probiotics as well as their extraction from processed dairy products are reviewed in this review.

Author details

Kishwer Fatima Sherwani* and Dil Ara Abbas Bukhari
Government College Univeristy Lahore, Pakistan

*Address all correspondence to: kishwerfatima900@gmail.com

References

[1] Fuller AFRC R. Probiotics in man and animals. Journal of Applied Bacteriology. 1989;**66**(5):365-366. DOI: 10.1111/j.1365-2672.1989.tb05105.x

[2] Tambekar DH, Bhutada SA. An evaluation of probiotic potential of Lactobacillus. From milk of domestic animals and commercial available probiotic preparations inprevention of enteric bacterial infections. Recent Research in Science and Technology. 2010;**2**(10):82-88

[3] König H, Fröhlich J. Lactic acid bacteria. In: König H, Unden G, Fröhlich J, editors. Biology of Microorganisms on Grapes, in Must and in Wine. Springer; 2009. pp. 3-29

[4] Ayeni FA, Sanchez B, Adeniyi BA, et al. Evaluation of the functional potential of Weissella and Lactobacillus isolates obtained from Nigerian traditional fermented foods and cow's intestine. International Journal of Food Microbiology. 2011;**147**(97-104):31

[5] Arbuckle WS. Ice Cream. 4th ed. New York: Springer Science Business Media; 1986. p. 453

[6] Akin MB, Akin MS, Kırmacı Z. Effects of inulin and sugar levels on the viability of yogurt and probiotic bacteria and the physical and sensory characteristics in probiotic ice-cream. Food Chemistry. 2007;**104**:93-99

[7] Leandro EDS, De Arau Jo EA, Da Conceic A˜OLL, et al. Survival of Lactobacillus delbrueckii UFV H2b20 in ice cream produced with different fat levels and after submission to stress acid and bile salts. Journal of Functional Foods. 2013;**5**(1):503-507

[8] Homayouni A, Azizi A, Javadi M, Mahdipour S, Ejtahed H. Factors influencing probiotic survival in ice cream: A review. International Journal of Dairy Science. 2012;**7**:1-10

[9] Afzaal M, Khan AU, Saeed F, Arshad MS, Khan MA, Saeed M, et al. Survival and stability of free and encapsulated probiotic bacteria under simulated gastrointestinal conditions and in ice cream. Food Science & Nutrition. 2020;**8**(3):1649-1656. DOI: 10.1002/fsn3.1451

[10] Zaeim D, Sarabi-Jamab M, Ghorani B, Kadkhodaee R, Liu W, Tromp RH. Microencapsulation of probiotics in multi-polysaccharide microcapsules by electro-hydrodynamic atomization and incorporation into ice-cream formulation. Food Structure. 2020;**25**:100147. DOI: 10.1016/j.foostr.2020.100147

[11] Ahmadi A, Milani E, Madadlou A, Mortazavi SA, Mokarram RR, Salarbashi D. Synbiotic yogurt-ice cream produced via incorporation of microencapsulated lactobacillus acidophilus (la-5) and fructooligosaccharide. Journal of Food Science and Technology. 2014;**51**(8):1568-1574. DOI: 10.1007/s13197-012-0679-y

[12] Arslan AA, Gocer EMC, Demir M, et al. Viability of *Lactobacillus acidophilus* ATCC 4356 incorporated into ice cream using three different methods. Dairy Science & Technology. 2016;**96**:477-487. DOI: 10.1007/s13594-016-0282-5

[13] Ashaolu TJ. A review on selection of fermentative microorganisms for functional foods and beverages: The production and future perspectives. International Journal of Food Science & Technology. 2019;**54**:2511-2519

[14] Silva, Diego Romajrio; Sardi, Janaina de Ca¡ssia Orlandi, , Pitangui, Nayla de Souza; Roque, Sindy Magri; Silva, Andraca Cristina Barbosa da; Rosalen, Pedro Luiz (2020). Probiotics as an alternative antimicrobial therapy: Current reality and future directions. Journal of Functional Foods, 73, 104080. DOI: 10.1016/j.jff.2020.104080

[15] Carr FJ, Chill D, Maida N. The lactic acid bacteria: A literature survey. Critical Reviews in Microbiology. 2002;**28**:281-370

[16] de Melo Pereira GV, de Oliveira Coelho B, Magalhaes Júnior AIV, Soccol CR. How to select a probiotic? A review and update of methodsand criteria. Biotechnology Advances. 2018;**36**:2060-2076

[17] Saad N, Delattre C, Urdaci M, Schmitter JM, Bressollier P. An overview of the last advances in probiotic and prebiotic field. LWT - Food Science and Technology. 2013;**50**(1):1-16. DOI: 10.1016/j.lwt.2012.05.014

[18] Hill C, Guarner F, Reid G, Gibson GR, Merenstein DJ, Pot B, et al. Expert consensus document: The International Scientific Association for Probiotics and Prebiotics consensus statement on the scope and appropriate use of the term probiotic. Nature Reviews Gastroenterology & Hepatology. 2014;**11**(8):506-514. DOI: 10.1038/nrgastro.2014.66

[19] Parker M, Zobrist S, Donahue C, Edick C, Mansen K, Nadjari MHZ, et al. Naturally fermented milk from northern Senegal: Bacterial community composition and probiotic enrichment with *Lactobacillus rhamnosus*. Frontiers in Microbiology. 2018;**9**:2218. DOI: 10.3389/fmicb.2018.02218

[20] Wang S, Zhu H, Lu C, Kang Z, Luo Y, Feng L, et al. Fermented milk supplemented with probiotics and prebiotics can effectively alter the intestinal microbiota and immunity of host animals. Journal of Dairy Science. 2012;**95**(9): 4813-4822. DOI: 10.3168/jds.2012-5426

[21] Kakisu, Emiliano; Irigoyen, Aurora; Torre, Paloma; , De Antoni, Graciela L; Abraham, Analía G (2011). Physicochemical, microbiological, and sensory profiles of fermented milk containing probiotic strains isolated from kefir. Journal of Dairy Research, 78(4), 456-463. doi:10.1017/s0022029911000653

[22] Kakisu E, Irigoyen A, Torre P, De Antoni GL, Abraham AG. Physicochemical, microbiological and sensory profiles of fermented milk containing probiotic strains isolated from kefir. Journal of Dairy Research. 2011;**78**(4):456-463. DOI: 10.1017/s0022029911000653

[23] Dan T, Chen H, Li T, Tian J, Ren W, Zhang H, et al. Influence of *Lactobacillus plantarum* P-8 on fermented milk flavor and storage stability. Frontiers in Microbiology. 2019;**9**:3133. DOI: 10.3389/fmicb.2018.03133

[24] Zareba D, Ziarno M, Scibisz I, Gawron J. The importance of volatile compound profile in the assessment of fermentation conducted by *Lactobacillus casei* DN-114 001. International Dairy Journal. 2014;**35**(1):11-14. DOI: 10.1016/j.idairyj.2013. 09.009

[25] Mater DDG, Bretigny L, Firmesse O, Flores MJ, Mogenet A, Bresson JL, et al. *Streptococcus thermophilus* and *Lactobacillus delbrueckii* subsp. *bulgaricus* survive gastrointestinal transit of healthy volunteers consuming yogurt. FEMS Microbiology Letters. 2005;**250**(2):185-187. DOI: 10.1016/j.femsle.2005.07.006

[26] Guarner F, Perdigon G, Corthier G, Salminen S, Koletzko B,

Morelli L. Should yoghurt cultures be considered probiotic? British Journal of Nutrition. 2005;**93**(6):783-786. DOI: 10.1079/bjn20051428

[27] Taghizadeh Moghaddam S, Javadi A, Matin AA. Reduction of bisphenol A by *Lactobacillus acidophilus* and *Lactobacillus plantarum* in yoghurt. International Journal of Dairy Technology. 2020;**73**(4):737-742. DOI: 10.1111/1471-0307.12706

[28] Zhang SS, Xu ZS, Qin LH, Kong J. Low-sugar yogurt making by the co-cultivation of *Lactobacillus plantarum* WCFS1 with yogurt starter cultures. Journal of Dairy Science. 2020;**103**(4):3045-3054. DOI: 10.3168/jds.2019-17347

[29] Pimpin L, Wu JHY, Haskelberg H, Del Gobbo L, Mozaffarian D. Is butter back? A systematic review and metaanalysis of butter consumption and risk of cardiovascular disease, diabetes, and total mortality. PLoS One. 2016;**11**(6):e0158118. DOI: 10.1371/journal.pone.0158118

[30] Aloğlu H, Öner Z. Assimilation of cholesterol in broth, cream, and butter by probiotic bacteria. European Journal of Lipid Science and Technology. 2006;**108**(9):709-713. DOI: 10.1002/ejlt.200600137

[31] Ekinci FY, Okur OD, Ertekin B, Guzel-Seydim Z. Effects of probiotic bacteria and oils on fatty acid profiles of cultured cream. European Journal of Lipid Science and Technology. 2008;**110**(3):216-224. DOI: 10.1002/ejlt.200700038

[32] Yilmaz-Ersan L. Fatty acid composition of cream fermented by probiotic bacteria. Mljekarstvo. 2013;**63**(3):132-139. DOI: 10.3382/ps.2012-02874

[33] Watkins C, Murphy K, Dempsey EM, Murphy BP, O'Toole PW, Ross RP, et al. The viability of probiotics in water, breast milk, and infant formula. European Journal of Pediatrics. 2018;**177**(6):867-870. DOI: 10.1007/s00431-018-3133-y

[34] Mbye M, Baig MA, AbuQamar SF, El-Tarabily KA, Obaid RS, Osaili TM, et al. Updates on understanding of probiotic lactic acid bacteria responses to environmental stresses and highlights on proteomic analyses. Comprehensive Reviews in Food Science and Food Safety. 2020;**19**(3):1110-1124. DOI: 10.1111/1541-4337.12554

[35] Lourens-Hattingh A, Viljoen BC. Yogurt as probiotic carrier food. International Dairy Journal. 2001;**11**(1-2):1-17. DOI: 10.1016/s0958-6946(01)00036-x

[36] Ibrahim SA, Carr JP. Viability of bifidobacteria in commercial yogurt products in North Carolina during refrigerated storage. International Journal of Dairy Technology. 2006;**59**(4):272-277. DOI: 10.1111/j.1471-0307.2006.00282.x

[37] Vinderola CG, Bailo N, Reinhemier JA. Survival of probiotic microflora in Argentinian yoghurts during refrigerated storage. Food Research International. 2000;**33**(2):97-102. DOI: 10.1016/s0963-9969(00)00011-9

[38] Talwalkar A, Kailasapathy K. A review of oxygen toxicity in probiotic yogurts: Influence on the survival of probiotic bacteria and protective techniques. Comprehensive Reviews in Food Science and Food Safety. 2004;**3**(3):117-124. DOI: 10.1111/j.1541-4337.2004.tb00061.x

[39] Miller CW, Nguyen MH, Rooney M, Kailasapathy K. The control of dissolved oxygen content in probiotic yoghurts by alternative packaging materials. Packaging Technology and Science. 2003;**16**(2):61-67. DOI: 10.1002/pts.612

[40] Meybodi NM, Mortazavian AM, Arab M, Nematollahi A. Probiotic viability in yoghurt: A review of influential factors. International Dairy Journal. 2020;**109**:104793. DOI: 10.1016/j.idairyj.2020.104793

[41] Dave RI, Shah NP. Effectiveness of ascorbic acid as an oxygen scavenger in improving viability of probiotic bacteria in yoghurts made with commercial starter cultures. International Dairy Journal. 1997;**7**(6-7):435-443. DOI: 10.1016/s0958-6946(97)00026-5

[42] Cruz AG, Castro WF, Faria JAF, Lollo PCB, Amaya Farfán J, Freitas MQ, et al. Probiotic yogurts manufactured with increased glucose oxidase levels: Postacidification, proteolytic patterns, survival of probiotic microorganisms, production of organic acid and aroma compounds. Journal of Dairy Science. 2012;**95**(5):2261-2269. DOI: 10.3168/jds.2011-4582

[43] Cruz AG, Feria JAF, Walter EHM, Andrade RR, Cavalcanti RN, Oliveira CAF, et al. Processing optimization of probiotic yogurt containing glucose oxidase using response surface methodology. Journal of Dairy Science. 2010;**93**(11):5059-5068. DOI: 10.3168/jds.2010-3336

[44] Corcoran BM, Stanton C, Fitzgerald GF, Ross RP. Survival of probiotic lactobacilli in acidic environments is enhanced in the presence of metabolizable sugars. Applied and Environmental Microbiology. 2005;**71**(6):3060-3067. DOI: 10.1128/AEM.71.6.3060-3067.2005

[45] Sanz Y. Ecological and functional implications of the acidadaptation ability of Bifidobacterium: A way of selecting improved problotic strains. International Dairy Journal. 2007;**17**(11):1284-1289. DOI: 10.1016/j.idairyj.2007.01.016

[46] Rashidinejad A, Bahrami A, Salara AR, Rezaei A, Mahdi S. Co-encapsulation of probiotics with prebiotics and their application in functional/synbiotic dairy products. Critical Reviews in Food Science and Nutrition. 2020. DOI: 10.1080/10408398.2020.1854169

[47] https://www.intechopen.com/chapters/50992

[48] Mahrous H, El-Kholy WM, Elsanhoty RM. Production of new synbiotic yoghurt with local probiotic isolate and oat and study its effect on mice. Journal of Advances in Dairy Research. 2014;**24**:1-7. DOI: 10.4172/2329-888X.1000121

[49] Weichselbaum E. Probiotics and health: A review of the evidence. Nutrition Bulletin. 2009;**34**(4):340-373. DOI: 10.1111/j.1467-3010.2009.01782.x

[50] Yerlikaya O. Starter cultures used in probiotic dairy product preparation and popular probiotic dairy drinks. Journal of Food Science and Technology. 2014;**34**(2):221-229. DOI: 10.1590/fst.2014.0050

[51] Zago M, Fornasari ME, Carminati D, et al. Characterization and probiotic potential of *Lactobacillus plantarum* strains isolated from cheeses. Food Microbiology. 2011;**28**:1033-1010

[52] Piewngam P et al. Composition of the intestinal microbiota in extended-spectrum β-lactamase-producing Enterobacteriaceae carriers and non-carriers in Thailand. International Journal of Antimicrobial Agents. 2019;**53**(4):435-441. DOI: 10.1016/j.ijantimicag.2018.12.006

[53] Piewngam P, Zheng Y, Nguyen TH, Dickey SW, Joo H-S, Villaruz AE, et al. Pathogen elimination by probiotic

Bacillus via signalling interference. Nature. 2018;**562**:532-537. DOI: 10.1038/s41586-018-0616-y

[54] Chung LK, Raffatellu M. Probiotic fengycins disagree with Staphylococcus aureus colonization. Cell Research. 2019;**29**:93-94

[55] Moraes MCC, Costa PJC, Segundo ASG, Peruzzo DC. Assessment of probiotic bacterial strains effect on *S. aureus* biofilm on titanium discs with treated surfaces. Revista de Odontologia da UNESP. 2019;**48**:e20190096

[56] Goderska K, Agudo Pena S, Alarcon T. Helicobacter pylori treatment: Antibiotics or probiotics. Applied Microbiology and Biotechnology. 2018;**102**:1-7

[57] Aiba Y et al. Anti-Helicobacter pylori activity of non-living, heat-killed form of lactobacilli including *Lactobacillus johnsonii* No.1088. FEMS Microbiology Letters. 2017;**364**(11). DOI: 10.1093/femsle/fnx102

[58] Lahtinen SJ et al. Specific Bifidobacterium strains isolated from elderly subjects inhibit growth of *Staphylococcus aureus*. International Journal of Food Microbiology. 2007;**117**(1):125-128. DOI: 10.1016/j.ijfoodmicro.2007.02.023

[59] Lazarenko L, Babenko L, Sichel LS, Pidgorskyi V, Mokrozub V, Voronkova O, et al. Antagonistic action of Lactobacilli and Bifidobacteria in relation to Staphylococcus aureus and their influence on the immune response in cases of intravaginal Staphylococcosis in mice. Probiotics and Antimicrobial Proteins. 2012;**4**(2):78-89. DOI: 10.1007/s12602-012-9093-z

[60] Liao H et al. Enhanced antifungal activity of bovine lactoferrin-producing probiotic Lactobacillus casei in the murine model of vulvovaginal candidiasis. BMC Microbiology. 2019;**19**(1). DOI: 10.1186/s12866-018-1370-x

[61] Mottin V, Suyenaga ES. An approach on the potential use of probiotics in the treatment of skin conditions: Acne and atopic dermatitis. International Journal of Dermatology. 2018;**57**(12):1425-1432. DOI: 10.1111/ijd.13972

[62] Bowe WP, Filip JC, DiRienzo JM, Volgina A, Margolis DJ. Inhibition of propionibacterium acnes by bacteriocin-like inhibitory substances (BLIS) produced by Streptococcus salivarius. Journal of Drugs in Dermatology. 2006;**5**(9):868-870

[63] Kober MM, Bowe WP. The effect of probiotics on immune regulation, acne, and photoaging. International Journal of Women's Dermatology. 2015;**1**(2):85-89. DOI: 10.1016/j.ijwd.2015.02.001

[64] Dinan TG, Cryan JF. Brain gut microbiota axis and mental health. Psychosomatic Medicine. 2017;**79**(8):920-926

[65] Verdino J. The third tier in treatment: Attending to the growing connection between gut health and emotional well-being. Health Psychol Open. 2017;**4**(2):2055102917724335. DOI: 10.1177/2055102917724335

[66] Cerdo T, Ruiz A, Suárez A, Campoy C. Probiotic, prebiotic, and brain development. Nutrients. 2017;**9**:1247. DOI: 10.3390/nu9111247

[67] Holzer P, Farzi A. Neuropeptides, and the microbiota-gut brain axis. In: Lyte M, Cryan J, editors. Microbial Endocrinology: The Microbiota-Gut-Brain axis in Health and Disease. Advances in Experimental Medicine and Biology. 1st ed. Vol. 817. New York: Springer; 2014. pp. 195-219

[68] Lammers KM, Brigidi P, Vitali B, Gionchetti P, Rizzello F, Caramelli E, et al. Immunomodulatory effects of probiotic bacteria DNA: IL-1 and IL-10 response in human peripheral blood mononuclear cells. FEMS Immunology & Medical Microbiology. 2003;**38**(2):165-172

[69] Di Giacinto C, Marinaro M, Sanchez M, Strober W, Boirivant M. Probiotics ameliorate recurrent Th1-mediated murine colitis by inducing IL-10 and IL-10- dependent TGF-β-bearing regulatory cells. The Journal of Immunology. 2005;**174**(6):3237-3246

[70] Ouwehand AC. Antiallergic effects of probiotics. The Journal of Nutrition. 2007;**137**(3):794S-797S

[71] Thomas DW, Greer FR. Probiotics and prebiotics in pediatrics. Pediatrics. 2010;**126**(6):1217-1231

[72] Manzoni P, Lista G, Gallo E, Marangione P, Priolo C, Fontana P, et al. Routine *Lactobacillus rhamnosus* GG administration in VLBW infants: A retrospective, 6-year cohort study. Early Human Development. 2011;**87**:S35-S38

[73] Garrigues C, Johansen E, Pedersen MB. Complete genome sequence of *Bifidobacterium animalis* subsp. lactis BB-12, a widely consumed probiotic strain. Journal of Bacteriology. 2010;**192**(9):2467-2468

[74] Saavedra JM, Abi-Hanna A, Moore N, Yolken RH. Long-term consumption of infant formulas containing live probiotic bacteria: Tolerance and safety. The American Journal of Clinical Nutrition. 2004;**79**(2):261-267

[75] Ouwehand AC, Tölkkö S, Salminen S. The effect of digestive enzymes on the adhesion of probiotic bacteria in vitro. Journal of Food Science. 2001;**66**(6):856-859. DOI: 10.1111/j.1365-2621.2001.tb15186.x

[76] Kent RM, Doherty SB. Probiotic bacteria in infant formula and follow-up formula: Microencapsulation using milk and pea proteins to improve microbiological quality. Food Research International. 2014;**64**:567-576. DOI: 10.1016/j.foodres.2014.07.029

[77] Sekine I, Yoshihara S, Homma N, Hirayama T, Tonozuka S. Effects of bifidobacteria containing milk (TM1) on chemiluminescence reaction of peripheral leukocytes and mean corpuscular volume of red blood cells-a possible role of bifidobacteria on activation of macrophages. Therapeutics (Japan). 1985;**14**(5):691-695

[78] Tavan E, Cayuela C, Antoine J-M, Cassand P. Antimutagenic activities of various lactic acid bacteria against food mutagens: Heterocyclic amines. Journal of Dairy Research. 2002;**69**(2):335-341. DOI: 10.1017/S002202990200540X

[79] Pool-Zobel BL, Neudecker C, Domizlaff SJ, Schillinger U, Rumney C, Moretti M, et al. Lactobacillus- and Bifidobacterium-mediated antigenotoxicity in the colon of rats. Nutrition and Cancer. 1996;**26**:365-380

[80] Patole SK, Rao SC, Keil AD, Nathan EA, Doherty DA, Simmer KN, et al. Benefits of Bifidobacterium breve M-16V supplementation in preterm neonates - A retrospective cohort study. PLOS ONE. 2016;**11**(3):e0150775. DOI: 10.1371/journal.pone.0150775

[81] Venturi A, Gionchetti P, Rizzello F, Johansson R, et al. Impact on the composition of the faecal flora by a new probiotic preparation: Preliminary data on maintenance treatment of patients with ulcerative colitis. Alimentary Pharmacology & Therapeutics. 1999;**13**:1103-1108. DOI: 10.1046/j.1365-2036.1999.00560.x

[82] Heyman M, Ménard S. Probiotic microorganisms: How they affect intestinal pathophysiology. Cellular and Molecular Life Sciences. 2002;**59**(7):1151-1165. DOI: 10.1007/s00018-002-8494-7

[83] Gionchetti P, Rizzello F, Venturi A, Campieri M. Probiotics in infective diarrhoea and inflammatory bowel diseases. Journal of Gastroenterology and Hepatology. 2000;**15**:489-493. DOI: 10.1046/j.1440-1746.2000.02162.x

[84] Francavilla R, Lionetti E, Castellaneta S, Ciruzzi F, Indrio F, Masciale A, et al. Randomised clinical trial: *Lactobacillus reuteri* DSM 17938 vs. placebo in children with acute diarrhea double-blind study. Alimentary Pharmacology & Therapeutics. 2012;**36**(4):363-369

[85] Indrio F, Riezzo G, Raimondi F, Bisceglia M, Filannino A, Cavallo L, et al. *Lactobacillus reuteri* accelerates gastric emptying and improves regurgitation in infants. European Journal of Clinical Investigation. 2011;**41**(4):417-422

[86] Witthuhn RC, Schoeman T, Britz TJ. Characterisation of the microbial population at different stages of kefir production and kefir grain mass cultivation. The International Dairy Journal. 2005;**15**:383-389. DOI: 10.1016/j.idairyj.2004.07.016

[87] McFarland LV. Meta-analysis of probiotics for the prevention of traveler's diarrhea. Travel Medicine and Infectious Disease. 2007;**5**:97-105. DOI: 10.1016/j.tmaid.2005.10.003

[88] Choi CH, Jo SY, Park HJ, Chang SK, Byeon JS, Myung SJ. A randomized, double-blind, placebo-controlled multicenter trial of *Saccharomyces boulardii* in irritable bowel syndrome: Effect on quality of life. Journal of Clinical Gastroenterology. 2011;**45**:679-683. DOI: 10.1097/MCG.0b013e318204593e

[89] Dugoua JJ, Machado M, Zhu X, Chen X, Koren G, Einarson TR. Probiotic safety in pregnancy: A systematic review and meta-analysis of randomized controlled trials of Lactobacillus, Bifidobacterium, and Saccharomyces spp. Journal of Obstetrics and Gynaecology Canada. 2009;**31**:542-552

[90] Chmielewska A, Szajewska H. Systematic review of randomized controlled trials: Probiotics for functional constipation. World Journal of Gastroenterology. 2010;**16**:69-75

[91] Behnsen J, Deriu E, Sassone-Corsi M, Raffatellu M. Probiotics: Properties, examples, and specific applications. Cold Spring Harbor Perspectives in Medicine. 2013;**3**. DOI: 10.1101/cshperspect.a010074

[92] George Kerry R et al. Benefaction of probiotics for human health: A review. Journal of Food and Drug Analysis. 2018. DOI: 10.1016/j.jfda.2018.01.002

[93] Tejero-Sarinena S, Barlow J, Costabile A, Gibson GR, Rowland I. Antipathogenic activity of probiotics against Salmonella typhimurium and *Clostridium difficile* in anaerobic batch culture systems: Is it due to synergies in probiotic mixtures or the specificity of single strains? Anaerobe. 2013;**24**:60e5

[94] Kareem KY, Ling FH, Chwen LT, Foong OM, Asmara SA. Inhibitory activity of postbiotic produced by strains of Lactobacillus plantarum using reconstituted media supplemented with inulin. Gut Pathogens. 2014;**6**:1e7

[95] Folkman J. Angiogenesis. Annual Review of Medicine. 2006;**57**(1):1-18. DOI: 10.1146/annurev.med.57.121304.131306

[96] Chen X, Yang G, Song J-H, Xu H, Li D, Goldsmith J, et al. Probiotic yeast inhibits VEGFR signaling and angiogenesis in intestinal inflammation. PLoS One. 2013;**8**:1e7

[97] Takeda S, Hidaka M, Yoshida H, Takeshita M, Kikuchi Y, Tsend-Ayush C, et al. Antiallergic activity of probiotics from Mongolian dairy products on type I allergy in mice and mode of antiallergic action. Journal of Functional Foods. 2014;**9**:60e9

[98] Vidya S, Thiruneelakandan G. Probiotic potentials of lactobacillus and its anti-cancer activity. International Journal of Current Research. 2015;7:20680e4

[99] Gayathri D, Rashmi BS. Anti-cancer properties of probiotics: A natural strategy for cancer prevention. EC Nutrition. 2016;**5**:1191e202

[100] Vafaeie F. Critical Review on Probiotics and its Effect on cancer. Vol. 2. Cancer Press; 2016. p. 30e4

[101] Kahouli I, Malhotra M, Alaoui-Jamali MA, Prakash S. In-vitro characterization of the anti-cancer activity of the probiotic bacterium *Lactobacillus fermentum* NCIMB 5221 and potential against colorectal cancer cells. Journal of Cancer Sciences and Therapy. 2015;7:224e35

[102] Awaisheh SS, Obeidat MM, Al-Tamimi HJ, Assaf AM, Elqudah JM, Al-khazaleh JM, et al. In vitro cytotoxic activity of probiotic bacterial cell extracts against Caco-2 and HRT-18 colorectal cancer cells. Milk Science International. 2016;**69**:27e31

[103] Xu K, Cai H, Shen Y, Ni Q, Chen Y, Hu SL, et al. Management of corona virus disease-19 (COVID-19): The Zhejiang experience. Zhejiang Da Xue Xue bao Yi Xue ban = Journal of Zhejiang University, Medical Science. 2020;**49**

[104] Al Kassaa I. New Insights on Antiviral Probiotics: From Research to Applications. 1st ed. Springer; 2017. DOI: 10.1007/978-3-319-49688-7

[105] Benson HJ. Microbiological Applications Laboratory Manual in General Microbiology. 8th ed. New York, NY: McGraw-Hill; 2002. pp. 58-60

[106] Amann RI, Ludwig W, Schleifer KH. Phylogenetic identification and in situ detection of individual cells without cultivation. Microbiological Reviews. 1995;**59**:143-169

[107] Vandamme P, Pot B, Gillis M, de Vos P, Kersters K, Swings J. Polyphasic taxonomy, a consensus approach to bacterial systematics. Microbiological Reviews. 1996;**60**:407-438

[108] de los Reyes-Gavilan CG, Suarez A, Fernandez-Garcıa M, et al. Adhesion of bile-adapted Bifidobacterium strains to the HT29-MTX cell line is modified after sequential gastrointestinal challenge simulated in vitro using human gastric and duodenal juices. Research in Microbiology. 2011;**162**:514-519

[109] Masco L, Crockaert C, van Hoorde K, et al. In vitro assessment of the gastrointestinal transit tolerance of taxonomic reference strains from human origin and probiotic product isolated of Bifidobacterium. Journal of Dairy Science. 2007;**90**:3572-3578

[110] Ward LJH, Timmins MJ. Differentiation of Lactobacillus casei, Lactobacillus paracasei and Lactobacillus rhamnosus by polymerase chain reaction. Letters in Applied Microbiology. 1999;**29**:90-92. DOI: 10.1046/j.1365-2672.1999.00586.x

[111] Vinderola CG, Prosello W, Ghiberto D, Reinheimer J. Viability of probiotic (*Bifidobacterium*, *Lactobacillus*

acidophilus and *Lactobacillus casei*) and non probiotic microflora in argentinian fresh cheese. Journal of Dairy Science. 2000;**83**:1905-1911

[112] Kasimoglu A, Goncuoglu M, Akgun S. Probiotic white cheese with *Lactobacillus acidophilus*. International Dairy Journal. 2004;**14**:1067-1073

[113] Phillips M, Kailasapathy K, Tran L. Viability of commercial probiotic cultures (L. acidophilus, Bifidobacterium sp., L. casei, L. paracasei and L. rhamnosus) in cheddar cheese. International Journal of Food Microbiology. 2006;**108**:276-280

[114] Liu C, Zhang Z-Y, Dong K, Yuan J-P, Guo X-K. Antibiotic resistance of probiotic strains of lactic acid bacteria isolated from marketed foods and drugs. Biomedical and Environmental Sciences. 2009;**22**(5):401-412. DOI: 10.1016/s0895-3988(10)60018-9

[115] Mishral AM, Sharma K. Isolation and characterization of probiotic microorganism from fermented dairy products. GERF Bulletin of Biosciences. 2014;**5**(1):10-14

[116] Tavakoli M, Hamidi-Esfahani Z, Hejazi MA, Azizi MH, Abbasi S. Characterization of probiotic abilities of Lactobacilli isolated from Iranian Koozeh Traditional Cheese. Polish Journal of Food and Nutrition Sciences. 2017;**67**(1). DOI: 10.1515/pjfns-2016-0003

[117] Sharpe ME. Identification of the lactic acid bacteria. In: Skinner FA, Lovelock DW, editors. Identification Methods for Microbiologist. London: Academic Press; 1979. pp. 233-259

[118] Masalam B, Maged S, Bahieldin A, Alharbi MG, Al-Masaudi S, Al-Jaouni SK, et al. Isolation, molecular characterization and probiotic potential of lactic acid bacteria in saudi raw and fermented milk. Evidence-Based Complementary and Alternative Medicine. 2018;**2018**:1-12. DOI: 10.1155/2018/7970463

[119] Qian Z, Zhao D, Yin Y, Zhu H, Chen D. Antibacterial activity of Lactobacillus strains isolated from Mongolian yogurt against *Gardnerella vaginalis*. BioMed Research International. 2020, 2020:1-9. DOI: 10.1155/2020/3548618

[120] Hajigholizadeh M, Mardani K, Moradi M, Jamshidi A. Molecular detection, phylogenetic analysis, and antibacterial performance of lactic acid bacteria isolated from traditional cheeses, North-West Iran. Food Science & Nutrition. 2020;**8**(11):6007-6013. DOI: 10.1002/fsn3.1887

[121] Koohestani M, Moradi M, Tajik H, Badali A. Effects of cell-free supernatant of *Lactobacillus acidophilus* LA5 and *Lactobacillus casei* 431 against planktonic form and biofilm of *Staphylococcus aureus*. Veterinary Research Forum. 2018;**9**:301-306

[122] Albano C, Morandi S, Silvetti T, Casiraghi MC, Manini F, Brasca M. Lactic acid bacteria with cholesterol-lowering properties for dairy applications: In vitro and in situ activity. Journal of Dairy Science, S0022030218308671. 2018. DOI: 10.3168/jds.2018-15096

[123] Ranadheera CS, Vidanarachchi JK, Rocha RS, Cruz AG, Ajlouni S. Probiotic delivery through fermentation: Dairy vs. non-dairy beverages. Fermentation. 2017;**3**:67

[124] Ness IF, Diep DB, Ike Y. Enterococcal bacteriocins and antimicrobial proteins that contribute to niche control. In: Gilmore MS, Clewell DB, Ike Y, Shankar N, editors. Enterococci: From

Commensals to Leading Causes of Drug Resistant Infection [Internet]. Boston, MA: Massachusetts Eye and Ear Infirmary; 2014

[125] Bakr SA. The potential applications of probiotics on dairy and non-dairy foods focusing on viability during storage. Biocatalysis and Agricultural Biotechnology. 2015;**4**:423-431

[126] Song S, Lee SJ, Park D-J, et al. The anti-allergic activity of Lactobacillus plantarum L67 and its application to yogurt. Journal of Dairy Research. 2016;**99**:9372-9382

[127] Wu C, He G, Zhang J. Physiological and proteomic analysis of Lactobacillus casei in response to acid adaptation. Journal of Industrial Microbiology and Biotechnology. 2014;**41**:1533-1540. DOI: 10.1007/s10295-014-1487-3